T0201523

CHINA AND TAIWAN ─────────

China Today

CHINA AND TAIWAN ———

Steven M. Goldstein

polity

First published in 2015 by Polity Press

Polity Press
65 Bridge Street
Cambridge CB2 1UR, UK

Polity Press
350 Main Street
Malden, MA 02148, USA

ISBN-13: 978-0-7456-5999-2
ISBN-13: 978-0-7456-6000-4 (pb)

A catalogue record for this book is available from the British Library.

Library of Congress Cataloging-in-Publication Data
Goldstein, Steven M.
    China and Taiwan / Steven Goldstein.
        pages   cm. – (China today)
    Includes bibliographical references and index.
    ISBN 978-0-7456-5999-2 (hardback : alk. paper) – ISBN 978-0-7456-6000-4
(pbk. : alk. paper)   1. China–Foreign relations–Taiwan.   2. Taiwan–Foreign
relations–China.   3. China–Foreign relations–United States.   4. United States–
Foreign relations–China.   5. Taiwan–Foreign relations–United States.   6. United
States–Foreign relations–Taiwan.   7. Taiwan–International status. I. Title.
    DS740.5.T28G65 2015
    327.51051249–dc23
                            2015010144

Typeset in 11.5 on 15 pt Adobe Jenson Pro
by Toppan Best-set Premedia Limited
Printed and bound in the UK by Clays Ltd St Ives PLC

For further information on Polity, visit our website: politybooks.com

Contents

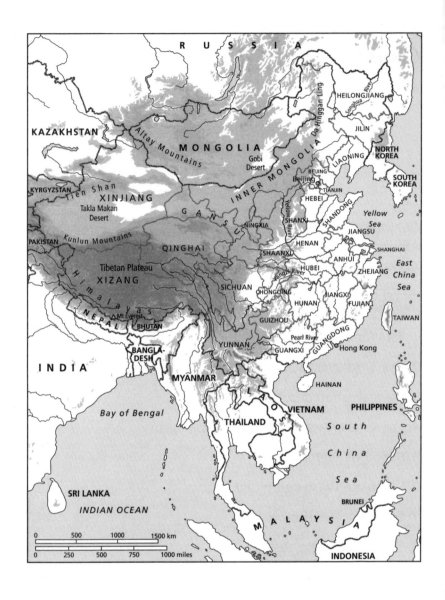

Chronology

1895	Taiwan becomes a Japanese colony by the Treaty of Shimonoseki
1911–12	Chinese republican revolution and fall of the Qing dynasty
1937–45	Anti-Japanese war
1943	Cairo Conference calls for Taiwan to be returned to China after the war
1945–9	Chinese Civil War between the Nationalists (KMT) and the Communists (CCP)
1945	Kuomintang troops accept the Japanese surrender on Taiwan
1947	February 28 uprising (2.28)
1949	Founding of the People's Republic of China (PRC); Kuomintang-dominated Republic of China moves to Taiwan
1950–3	Korean War; Truman orders US Seventh Fleet "to prevent any attack on Taiwan" and calls for the ROC to stop operations against the mainland
1953–7	First Five-Year Plan: the PRC adopts Soviet-style economic planning
1954	Constitution of the PRC implemented: first meeting of the National People's Congress; the US signs Mutual Defence Treaty with the ROC

1954–5	First Taiwan Strait crisis
1957	Hundred Flowers movement: brief period of political debate followed by repressive anti-rightist movement
1958	Second Taiwan Strait crisis
1958–60	Great Leap Forward: Chinese Communist Party aims to transform the agrarian economy through rapid industrialization and collectivization
1959	Tibetan uprising and the departure of the Dalai Lama for India
1959–61	Three years of natural disasters: widespread famine, with millions of deaths resulting largely from the policies of the Great Leap Forward
1960	"Sino-Soviet split"
1962	Sino-Indian border skirmishes
1964	First PRC atom bomb detonation
1971	UN General Assembly votes to replace the ROC with the People's Republic of China as representative of "China"
1966–76	Great Proletarian Cultural Revolution
February 1972	"Shanghai Communiqué," issued during Richard Nixon's visit to China, pledges that neither the US nor China will "seek hegemony in the Asia-Pacific region"
April 1975	Death of Chiang Kai-shek
July 1976	The Great Tangshan Earthquake: by death toll, the largest earthquake of the twentieth century
September 1976	Death of Mao Zedong
October 1976	Ultra-leftist Gang of Four removed from leadership
1978–89	Democracy Wall movement

1978	Beginning of Chinese economic reform and openness
1978	Introduction of one-child policy restricting married urban couples to one child
1979	Diplomatic relations established between the US and the PRC and broken with the ROC; US Congress passes Taiwan Relations Act
1979	PRC invades Vietnam
1982	US and PRC sign arms sales communiqué
December 1984	Margaret Thatcher co-signs Sino-British Joint Declaration agreeing to transfer sovereignty over Hong Kong to the PRC in 1997
1986	Democratic Progressive Party founded
January 1988	Chiang Ching-kuo dies and is succeeded as president of the ROC by Lee Teng-hui
1989	Tiananmen Square movement and crackdown
1989–2002	Jiang Zemin serves as general secretary of the Chinese Communist Party and president of the PRC
1991	Period of national mobilization for suppression of the communist rebellion ended
1992	Deng Xiaoping's southern inspection tour restarts process of economic reform and development; ARATS and SEF meet in Singapore
1995	Lee Teng-hui visits the United States
1996	Mainland conducts missile tests during Taiwan elections; US sends two aircraft carrier groups to the area; Lee Teng-hui elected president of the ROC
May 1999	US bombing of Chinese embassy in Belgrade
1999	Falun Gong demonstrations in Beijing

2000	DPP candidate Chen Shui-bian elected president of the ROC
2001	China joins World Trade Organization
2002	Taiwan joins the World Trade Organization as the "Separate Customs Territory of Taiwan, Penghu, Kinmen and Matsu (Chinese Taipei)"
2002–12	Hu Jintao serves as general secretary of the Chinese Communist Party and president of the PRC
2002	SARS outbreak
2004	Chen Shui-bian re-elected president of the ROC
2005	China passes Anti-Succession Law
2007	China overtakes the US as the world's biggest emitter of CO_2
2008	Sichuan earthquake; Kuomintang candidate Ma Ying-jeou elected president of the ROC; Hu Jintao announces six points for managing relations with Taiwan
2008	Summer Olympic Games held in Beijing
2010	Shanghai World Exposition
2012	Xi Jinping elected general secretary of the CCP (and president of PRC from 2013); Ma Ying-jeou re-elected president of the ROC

Abbreviations

ARATS	Association for Relations Across the Taiwan Straits
ASEAN	Association of South-East Asian Nations
CCP	Chinese Communist Party
DPP	Democratic Progressive Party
ECFA	Economic Cooperation Framework Agreement
GDP	Gross domestic product
KMT	Kuomintang (Nationalist Party)
NPC	National People's Congress
PLA	People's Liberation Army
PRC	People's Republic of China
ROC	Republic of China
SEF	Straits Exchange Foundation
TPP	Trans-Pacific Partnership
TRA	Taiwan Relations Act
TSEA	Taiwan Security Enhancement Act
WTO	World Trade Organization

Men make their own history, but they do not make it as they please; they do not make it under self-selected circumstances, but under circumstances existing already, given and transmitted from the past.

Karl Marx, *The Eighteenth Brumaire of Louis Napoleon*

For Erika, without whom...

Acknowledgments

Academics, like professional gamblers, usually accumulate a long string of debts. I have been no exception. My study of Taiwan came relatively late in my academic career and I have benefited from intellectual inspiration and challenges coming from many sources. My students at Smith College, especially the recent influx of students from China, have provided not only research assistance but questions and comments that have forced me to clarify my thinking on cross-strait relations.

I have also been fortunate to have led delegations from the Taiwan Studies Workshop of the Fairbank Center for Chinese Studies at Harvard University that have travelled to Taiwan and the mainland for more than a decade. We have met (and continue to meet) with academics and government officials to discuss cross-strait relations and American foreign policy. Our interlocutors (too numerous to mention) on both sides have been generous with their time as well as their willingness to discuss some very difficult questions with candour and, often, good humour.

However, special recognition must go to the members of the delegation who, year after year, left their families right after New Year to take part in our expeditions. On plane rides as well as in restaurants and hotel bars, I learned an enormous amount from Tom Christensen, Joe Fewsmith, Taylor Fravel, Sheena Chestnut Greitens, Iain Johnston, Robert Ross, and the late Alan Wachman. However, there is one member of the group who, I am confident, we would all agree deserves special mention – Alan Romberg. With his encyclopedic and precise knowledge of cross-strait relations and American foreign policy, he was the one whom we consistently turned to for wisdom and guidance on difficult or arcane questions. As this book demonstrates, he has had an incalculable influence on my thinking about Taiwan. I couldn't be more grateful. However, to preserve his good name as well as those of my fellow travellers to China, I have quickly to add that they are in no way responsible for this work.

Final mention has to go to those who helped enhance the quality and coherence of the discussion which follows. Pascal Porcheron and Louise Knight at Polity Press were patient when I failed to meet deadlines and helped to sharpen my argument. Saikun Shi provided research assistance. Samantha Wood and Caroline Richmond were amazing editors who performed magic on this manuscript.

Introduction

For more than six decades, the embers of the post-World War II conflict between Taiwan and the mainland of China have threatened to
burst into flames, engulfing the Taiwan Strait in a war that could
quickly become a broader and more dangerous conflict between the
United States and China.

The roots of today's cross-strait tensions are relatively straightfor-
ward. In 1949, after driving the government of the Republic of China
(ROC) – often referred to as the "Nationalists" – off the mainland and
onto the island of Taiwan, the Chinese Communist Party (CCP)
declared the establishment of the People's Republic of China (PRC).
Today, that government in Beijing considers itself the legitimate ruler
of all China, including Taiwan. It views the continued separation of
the island from the mainland, as well as its governance by another
political authority claiming equal sovereignty, as preventing both
closure in the civil war and restoration of the full territorial integrity
of the Chinese nation. Although PRC leaders have committed to
peaceful modes of achieving this reunification as their preference, they
nevertheless retain the option to use force in their efforts to incorporate
the island into China.

The authorities on Taiwan, on the other hand, insist that, regardless
of the defeat on the mainland in 1949, it remains the same government
that ruled China before the forced relocation. For some of the period
after its defeat, the ROC claimed to govern all of China despite the
mainland's occupation by "communist bandits." Today, decades after

the major world powers (including the United States) finally recognized the PRC as the legitimate government of China, the government on Taiwan continues to assert the ROC's status as a sovereign and independent state on the international stage. Although economic relations with the mainland have flourished, the ROC has resisted discussions aimed at resolving cross-strait political and military disagreements.

This brief account of the origins of cross-strait relations tells only part of the story. The United States became entangled in China's internal politics during World War II and has remained so since, despite several efforts at disassociation. Washington backed the Kuomintang (KMT or Nationalist Party) during the civil war with the Communist Party and, today, remains the ultimate guarantor of the safety of Taiwan. Communist leaders have, since 1949, considered the United States to be the principal obstacle to the incorporation of the island into the new Chinese state.

Thus the policies of both China and Taiwan have, of necessity, been focused on the United States even as Washington has had, in turn, to consider its objectives in regard to each side of the strait in formulating policy toward the area. For this reason, the discussion that follows proceeds from the premise that cross-strait relations cannot be fully understood if the focus is simply on the bilateral relationship between the two sides in an earlier domestic conflict. Despite their origins in the Chinese Civil War, these relations have had, from their beginning, a significant international dimension as a result of continued American involvement and, as a result, have taken on a triangular pattern.

For more than sixty years, this triangular configuration has remained the defining characteristic of cross-strait relations. However, it has not been a static configuration. The triangle has evolved over time in response to the policies of the three actors as well as to the broader international environment. These policies, which have accumulated over more than six decades, have created perceptions, assumptions, and commitments that together are the foundation of the present triangle in the Taiwan Strait. As is the case in so many other global hotspots, the past weighs heavily on the present and continues to shape interactions.

To assess the influence of the past on the contemporary situation, the analysis that follows posits that relations have gone through two distinct configurations since the end of World War II. These stages of development, despite their very different natures, combine to have a profound impact on the current policy in the area. The two periods are separated by the decade of the 1990s, with the most dramatic single event marking the passage into a new era in cross-strait relations being the end of KMT authoritarian rule and the emergence of democracy in Taiwan.

Before democratization, Taiwan's mainland policy was made by a small group of KMT leaders who were preoccupied with regaining power on the mainland and who treated Taiwan simply as a provincial jumping-off point for realizing that larger ambition. For them, the Taiwan Strait was still the front line in a continuing civil war. There were, to be sure, sporadic secret contacts between the two sides. However, aside from occasional military forays and frequent public propaganda statements across the strait, there were no interactions between the two sides that could be considered "relations." To the extent that there were any "relations," they were manifested in Sino-American dealings over the status of Taiwan, which was rooted in the post-war controversies that periodically flared into crises in the area. These were the years of the Cold War and the "Red Scare." China was viewed by Washington as the spear point of the international communist movement in Asia. By the mid-1950s the United States not only refused to acknowledge the communist victory in the civil war, as manifested by its continued recognition of the ROC as the government of China, but also denied that the PRC had sovereignty over Taiwan.

Until 1972, the United States was in the middle of the cross-strait dispute. American policy in the area was one of dual deterrence (for this term, see Bush 2005). Washington sought to prevent the Kuomintang on Taiwan from provoking a clash with the mainland that would drag it into a war with Beijing, while at the same time deterring a possible mainland attack on the island by its military presence. The United States engaged the mainland in 136 sessions of ambassadorial talks intended to de-escalate tensions in the area. China, however, would have none of it. Like the KMT on Taiwan, it regarded the cross-strait conflict as a domestic matter and American interference as a violation of its newly won sovereignty. Taiwan's status was considered a matter to be settled by the two sides themselves, and Beijing's representatives consistently argued that China would accept nothing less than American abandonment of Taiwan – an unlikely step given the political environment in the United States. It was a dialogue of the deaf.

It was against the background of this Sino-American deadlock over the status of Taiwan that the rapprochement of the 1970s, beginning with the visit of President Nixon and mutual recognition in the Carter administration, took place. As we shall see, the Sino-American differences over Taiwan proved no more soluble than they had been earlier, and differences nearly wrecked the process of normalization. However, both sides sought a better rapport, and, by means of ambiguous statements, muted disagreements or simple papering over the still sharp divisions over Taiwan, Sino-American relations went ahead into the 1990s – and into a new stage in the triangular relationship.

As noted earlier, it was the democratization of Taiwan and the end of KMT authoritarian rule during the 1990s that was the occasion for the transition to this new stage in cross-strait relations. This action enfranchised a portion of the Taiwan population who had lived on the island before World War II and whose orientation toward the island

and its relation to the mainland was fundamentally different from what had previously been official policy.

The roots of this new orientation and the subsequent shift in Taiwan's policy that resulted were in the past. Specifically, they were the result of a unique historic relationship between Taiwan and the mainland of China as well as the impact of the period of KMT authoritarian rule. Until the mid-seventeenth century, when it finally became a minor subdivision of the Chinese empire, the island was better known to pirates in the area than it was to the rulers of China. After two centuries of neglect by the mainland, Taiwan finally achieved provincial status. However, after less than a decade it was ceded to the Japanese empire in 1895, following China's defeat in the Sino-Japanese War, and became Tokyo's first colony. It retained that status for fifty years, until Japan's surrender at the end of World War II.

The distant relationship with imperial China and, more importantly, Taiwan's half-century as a Japanese colony would play a central role in shaping the domestic political environment on the island. When the Kuomintang army arrived to reclaim Taiwan after the war, it encountered an ethnically Chinese population that appeared to be more Japanese than Chinese and that shared little of the antipathy to Japan felt by the arriving mainlanders, who had just endured eight years of brutal occupation and war. The clash of cultures and history between the arrivals from the mainland (known as *waisheng*, or those from outside the province) and the Chinese whose ancestors had come before the end of the war (known as *bensheng* – those from within the province) became immediately apparent, and tensions grew, leading to an armed confrontation in 1947. The brutal suppression of local activists by mainland troops marked the end of any hope of greater self-rule for the islanders and initiated an authoritarian ROC government dominated by the newly arrived Kuomintang.

This cleavage between these two populations came to define Taiwan politics for more than four decades, as the mainlander government moved to reshape the island to meet its needs in the civil war against its enemy across the strait. Taiwan was subjected to what amounted to military rule, which suspended the constitution and virtually excluded the local population from political participation except at the very local level. In an effort to rally the population around the cause of retaking the mainland, the KMT government sought to "Sinify" the local population by imposing mainland values, history, and language to replace those associated with Taiwan. The result of these policies was that, over time, much of the resistance to authoritarian rule came to be associated with the *bensheng* population, who, in reaction to the forced Sinification, fashioned the island's past into a narrative that, contrary to the official policy, emphasized its distinctive history and identity as well as its extended separation from the mainland.

Thus, with democratization, mainland policy became subject to the influence of a population that had already become deeply divided over the question of identity during the previous period. The unique history of the island and the experience of mainland rule under KMT auspices had engendered a search for a distinctively Taiwanese identity, and the nature of the relationship with the mainland became a contested political issue. In contrast to the previous period, when the island's relationship with the mainland was taken as a given, it became, by the end of the twentieth century, an issue considered subject to negotiation among equals, with the newly formed opposition party floating the idea of independence.

However, democratization on Taiwan not only led to a questioning of the assumption regarding the island's status as a part of China, it also saw the end of the earlier refusal of the government to have any contacts with the "enemies" on the mainland. Democracy empowered the business community, and, with commercial interests leading the way, contact rather than conflict between the two sides became the dominant theme in cross-strait relations. In 1992, this new stage in the relationship was marked by a meeting between unofficial organizations from the two sides – the first since 1945.

In short, Taiwan's earlier policies of hostility and refusal to allow any contact with the mainland – outside of military provocations promoted by a bitter, defeated KMT leadership – were ended. The foundations of the unprecedented, multifaceted relationship in trade, investment, tourism, and official consultations that characterize contemporary cross-strait relations were laid. However, this policy was now subject to the pressures of an electorate far more ambivalent about the political nature of the relationship and clearly reluctant to replace the rule of one mainland government with another.

Taiwan's democratization had also shifted the central focus of cross-strait relations away from Sino-American diplomacy. One scholar (Su 2009) has referred to the period after the 1990s as "a tail wagging two dogs." After the 1990s, the United States and China were forced to adjust to policies resulting from domestic politics in Taiwan over which they had very little control and which were increasingly coming to shape the triangular configuration of relations (Chu and Nathan 2007–8).

For the mainland, the result was that the management of cross-strait relations became dramatically more complicated. The relationship with Taiwan that developed after the 1990s was a multifaceted one that encompassed a wide range of issues, including investment, culture, tourist exchanges, and governmental agreements. It operated on many levels, involving individual citizens, party members, and government officials. Most challenging for the mainland were the domestic political currents on Taiwan, which often pushed the limits of Beijing's long-established policies regarding the island's relationship with the mainland. Fundamental principles laid down by the mainland in the previous period were proving ill-suited to the new environment.

The same could be said for China's relationship with the United States. Beijing's frustration in managing an increasingly complex cross-strait relationship often caused it to look to the United States as either a cause of, or a solution to, its problems. The distrust of American motives rooted in the previous period remained. They had been neither dispelled nor, more importantly, solved by the earlier ambiguous agreements. This threatened at times to disrupt Sino-American relations, while at others Beijing looked to Washington to cooperate in limiting provocative behavior on the part of the newly democratic Taiwan.

The new period in cross-strait relations posed challenges for the United States as well. In some respects these were not new challenges. In the period after recognition, domestic political pressures and concerns for the American image in Asia had required that a delicate balance be maintained between enhancing the post-Cold War relationship with China and appearing not to abandon Taiwan. After democratization on Taiwan this balance was complicated, on the one hand, by a Chinese military build-up in response to the uncertain direction of cross-strait relations and, on the other, by the possibility that provocative policies resulting from Taiwan's new democratic politics would trigger a mainland response.

American policy became once more one of dual deterrence – only this time China was a far more formidable opponent and a democratic Taiwan more difficult to restrain. Moreover, in the previous period the provocation on Taiwan's part that might drag the United States into a war was military in nature. Now the principal danger was that provocative actions resulting from the island's domestic politics would be the cause of a military response from the mainland that could involve the United States.

In short, today the United States remains very much in the middle of the cross-strait relationship. Its options in playing its role in the reconfigured triangle, like those of the mainland, are constrained by past policies and perceptions. Moreover, in dealing with the two sides, Washington is bound by ambiguous commitments and agreements of the earlier period. This only increases the difficulty of its position.

The analysis that follows is intended to demonstrate the propositions regarding cross-strait relations outlined above. It is divided into two sections: the first is focused on the period before the 1990s and the second on the period afterward. Each is organized differently. The first proceeds in a largely chronological manner, tracing the evolution of cross-strait relations as an issue between the United States and China as well as the nature of the policies pursued by the mainland and Taiwan. The second section covers contemporary relations, and its organization is thematic. In individual chapters it discusses the way in which democratization in the presidencies of Chen Shui-bian (2000–8) and Ma Ying-jeou (2008–) has shaped Taiwan's policies and the manner in which the United States and China have tried to cope with those policies as well as each other's reactions to them. The analysis then turns to the two most important issues in contemporary relations – trade and investment and the military balance. A conclusion discusses the current state of cross-strait relations in light of the patterns of the past and speculates on the parameters of future developments.

1 | An Island of Unsettled Status

On February 27, 1947 – a year after Japan's formal surrender of its Taiwanese colony – agents of the mainland Republic of China government seized the cigarettes and cash of a street vendor in the capital city of Taipei and beat her. Facing an angry crowd, the agents claimed that the vendor was violating the official Tobacco Monopoly by selling untaxed cigarettes. During the ensuing confrontation, a bystander was shot and killed by an ROC agent. The following day, an estimated 2,000 demonstrators marched first to the headquarters of the Tobacco Monopoly and then on to the office of the governor, General Chen Yi, where security officers fired into the crowd.

These events sparked anti-government demonstrations across the island. Governor Chen entered into talks with the popular opposition, which quickly escalated from an investigation of the two shootings to a call for governmental reform. These tentative negotiations ended on March 8, when ROC reinforcements landed on the island. Over the next weeks, many of Taiwan's intellectual and political elite were methodically killed or arrested, while the general populace faced random killings and other atrocities. The total number of deaths during this period has been estimated to be between 10,000 and 20,000. These events became collectively known as the "February 28 Incident," or simply 2.28, and would play a decisive role in shaping the subsequent political development of Taiwan, as well as its relationship with the mainland. To fully understand the incident's profound impact, we must delve further into the history of Taiwan.

Harsh governmental response to popular demonstrations was a familiar reality in post-war China. The 2.28 Incident was not the first time that the Kuomintang-dominated Republic of China government had encountered resistance as it sought to re-establish its authority in areas previously occupied by the Japanese. Such episodes escalated to eventually become part of the civil war that raged on the mainland from 1947 to 1949. Yet the unrest on Taiwan was distinct in many ways from other opposition faced by the KMT government.

Unlike the contested mainland areas, Taiwan had not been a victim of Japanese wartime occupation. The island was a Japanese colony – its first and oldest colony, in fact – that had been ceded to Tokyo by the terms of the treaty that ended the Sino-Japanese War of 1895. Taiwan occupied an anomalous position during the early twentieth century, which was a period of emerging Chinese nationalism and determination to regain full national sovereignty. Sun Yat-sen, founder of the KMT, paid it scant attention. It was only in the early 1940s that his successor, Chiang Kai-shek, began to demand the island's return to China. Mao Zedong, leader of the Chinese Communist Party, adopted a similar position, although earlier he had argued that Taiwan, like Korea, should become an independent country after the war (Wachman 2007: chs 4 and 5).

The ambivalence of China's twentieth-century politicians regarding the status of Taiwan had historic provenance.[1] Until the end of the seventeenth century, the island was better known to Dutch and Spanish settlers, to Japanese pirates, and to the occasional Chinese civilian who came to the island in search of natural resources than it was to mainland officialdom. The first written report of a mainlander's visit was published in 1603. It was not until 1684 that a reluctant emperor agreed to the incorporation of what he called a "ball of mud...[of] no consequence" and Taiwan began appearing on mainland maps. In the eyes of most mainland Chinese, Taiwan was a "wilderness...beyond the seas...populated by savages" (Teng 2004: 31–59).

This incorporation into the Chinese administrative structure came after the Qing dynasty defeated rebels supporting the previous dynasty who had used the island as a base. The court official Shi Lang convinced a sceptical emperor to finally pay attention to the island, arguing for its strategic importance in defending the Chinese coast from foreign invasion as well as for its potential economic value. The expansionist Qing dynasty (responsible for nearly doubling the size of China) incorporated Taiwan as a mere sub-jurisdiction under the coastal province of Fujian. For the next two centuries, the imperial court ruled the troublesome island with a light hand, leaving much of its governance up to prominent Chinese families who had settled there.

In 1887, when it became apparent that Taiwan was coveted by foreign powers, the island gained provincial status and for the first time benefited from central government efforts to develop its economy. However, after China's defeat in the Sino-Japanese War of 1895 (fought nowhere near Taiwan), it was ceded to Japan in the Treaty of Shimonoseki. It would remain a colony until Tokyo's surrender at the end of World War II.

In 1895, Taiwan's population (probably around 3 million) consisted predominantly of ethnic Chinese: a group known as the Hoklo, who had migrated from coastal Fujian (c. 70 percent of the population), and the Hakka, or "guest people," who had come from northern China before making the crossing (c. 15–20 percent of the population). There were also a small number of native inhabitants of uncertain origin.[2]

Over the next five decades, Japan developed the island's economy to serve its needs.[3] At first, the emphasis was on developing agriculture (predominantly rice and sugar) and then, as war approached, on industrial development. Japan engaged in extensive construction of transportation systems and industrial infrastructure. By the end of the war, the island had developed a considerable manufacturing base (30 percent of GDP) and boasted a level of economic development second only to Japan's among Asian economies as well as a per capita income twice that of mainland China. Social transformation accompanied economic development.

The urban population – primarily factory workers – and the number of Taiwanese in the educational system grew. Although Japanese students were advantaged, some Chinese did advance to university, either in Japan or at the university established in Taipei by the colonial regime. These educational reforms were part of a broader program that sought to introduce the Chinese population of Taiwan to Japanese language and culture. By the early 1940s an estimated 58 percent of the island's inhabitants were functionally literate in Japanese, while 10 percent had actually adopted Japanese names. Among themselves, the Chinese population continued to speak the regional dialects of the mainland from which they had come. However, by the end of the war, Chinese language had been banned in newspapers as well as in schools, and as a result few islanders could speak the northern dialect that had become the national language on the mainland.

The political rule of the Japanese was strict and favored its own. However, it was generally honest, efficient, and based in law. Although the colonial government controlled the island's central administration, almost half the bureaucracy was Taiwanese, with low-level officials often elected by the local population. The main demand among the Taiwanese was for greater self-rule as a part of the empire. Only the Taiwanese Communist Party (a very minor player) spoke of independence. Yet the Japanese colonial authorities made only symbolic gestures to meet these demands. With the outbreak of World War II, military governors pushed the Taiwanese to assimilate into imperial culture. More than 200,000 Taiwanese (including a future president, Lee Teng-hui) served in the Japanese army. Finally, in a desperate attempt to gain the support of the island's population as defeat neared, Taiwan was promised full integration into the Japanese empire.

Japan surrendered to the Allies in August of 1945. Over the next few months, ROC soldiers and officials arrived on Taiwan from the mainland. Seven years at war with a hated and brutal Japanese enemy had left a devastated, demoralized China. When the ROC officials came to the island, they found a comparatively undamaged and prosperous economy, as well as a population that had been spared the suffering experienced on the mainland. Moreover, the islanders seemed more Japanese than Chinese: they spoke Japanese, dressed like Japanese, ate Japanese food, and, in some cases, had Japanese names. General Keh King-en, chief of the first ROC military mission, articulated the contemptuous attitude shared by many mainlanders when he characterized the island as a "degraded territory" and its inhabitants a "degraded people." To Keh and other mainlanders, Taiwan was "beyond the passes" – beyond the pale of true Chinese civilization (Kerr 1965: 72).

Although the islanders initially greeted the mainlanders, they soon reciprocated their disdain. Their "liberators" were a rag-tag army of often ignorant, undisciplined recruits. Within a year of the ROC's arrival, it became clear that the new government had none of Japan's efficiency and would deny Taiwanese aspirations for self-rule. The same corrupt, authoritarian management that was eroding the influence of the Nationalist government on the mainland was at work in Taiwan. Islanders began equating ROC "liberators" with Japanese occupiers, noting that the new occupiers lacked the competence of the former colonial regime. One contemporary observer notes that the prevalence of the phrase "Dogs go and pigs come" reflected not only Taiwanese contempt for corrupt mainlander rulers but also the fact that the ROC was being judged against the standard set by the Japanese – and found wanting (Kerr 1965: 97). This unfavorable comparison only heightened the disdain directed at the Taiwanese by the recently embattled mainlanders, who hated Japan and all things Japanese. On the eve of the 2.28 Incident, the ROC announced that the Chinese constitution due to go into effect at the end of 1947 would not extend to Taiwan. Governor Chen Yi claimed that, while mainland China was "advanced enough to enjoy the privileges of constitutional government," Taiwan was not (ibid.: 240).

During the February 1947 violence, representatives of the population spoke in terms of the rights and grievances of the "Formosans," claiming that islanders were entitled to the same treatment as "Chinese" (Kerr 1965: 278–90; Hsiau 2000; 57). Their plea was denied, and an official report on 2.28 spoke of a population divided by identity, of a people who desired to be Japanese and whose colonial education had engendered a "slave mentality" that had purged all knowledge of "the motherland" (Wang 2004: 6–8).

Facing an increasingly tenuous Nationalist position on the mainland, more than 2 million refugees migrated to Taiwan – including many military personnel and civilian administrators. As defeat loomed, rule on Taiwan became more authoritarian: in May of 1949, martial law was imposed on the island to defend against the communist threat. Articles of the nominally democratic ROC constitution passed on the mainland two years previously were suspended by the "Temporary Provisions and Special Legislation During the Period of Mobilization and Combating Rebellion," also in response to the conflict with the communists (Tien 1989: 105–11). In December, President Chiang Kai-shek and his ROC government officially moved to the island, establishing a new national capital in Taipei and designating Taiwan a province under the government that still claimed to rule all of China.

It soon became apparent that post-war Nationalist policies would focus on two objectives: justifying the ROC's claim as the legitimate government of all China and preparing to return to the mainland. The latter objective became the rationale behind the establishment of an autocratic regime in which the major government, military, and KMT party posts were dominated by newly arrived mainlanders (about 15 percent of the island's population). The parliament elected on the mainland became a mere rubber stamp, and political power was concentrated in the president's office and implemented by the police and military. Yet the parliament still sat with members representing all of China's provinces. The parliament symbolized not only the ROC's claim to national rule but also the inferior status of the Taiwanese – the only citizens actually ruled by the ROC – who were granted only token membership.

Under the pretext of crisis, this regime punished any individual or group that ostensibly challenged the legitimacy of KMT rule on Taiwan or its goal of recovering the mainland. While there was some room for political activity at the local level, evidence of actions hinting at Taiwanese nationalism or government opposition was frequently conflated with communist sympathies, and was dealt with harshly. This suppression became known as the "white terror." At the same time, the mainlander government sought to diminish any possibility of popular nationalism through a campaign of "de-Japanization and Sinicization," which was intended to address the assumed loss of Chinese identity on Taiwan, correct the assumed attitude of slavery to Japan, and gain the islanders' commitment to the mainland ROC government. Mandarin became the official language, required for government employment and in classroom instruction. In schools, the curriculum stressed historic ties with China, promoted traditional Chinese values, and spread Kuomintang anti-communist propaganda. At the same time, schools were pushed to ignore the complex history of Taiwan and to present Japanese colonialism on the island as indistinguishable from the empire's brutal conduct on the mainland (Hsiau 2000: ch. 6).

The distinction between *waisheng ren* (people from outside the province) and *bensheng ren* (people from the province) had become a fundamental political and cultural cleavage in post-war Taiwan. The former called the mainland home and considered the island and its people a resource for achieving homecoming. The latter were forced to accept a set of cultural values and a political system that confirmed their separate, inferior status. The post-war government on Taiwan was established with the basic principle of national identity deeply contested.

THE INTERNATIONAL DIMENSION: AN UNSETTLED STATUS

Five decades of Japanese colonial rule not only helped to create the deep cultural fissure that would define domestic Taiwan politics in the years after 1949. It also complicated efforts to define the relationship between the island of Taiwan and mainland China.

As discussed previously, in the treaty of 1895 following the First Sino-Japanese War, Taiwan was ceded in perpetuity to Japan. Since the island was not considered a part of Japanese-occupied China, its future status depended on a separate decision by the victorious allies. By the early 1940s, Chiang Kai-shek had abandoned the KMT's previous indifference to Taiwan's status and demanded that the island be returned to China. His demand carried weight, especially in Washington. President Roosevelt viewed China as both an essential wartime ally and a "strategic site" for maintaining peace in the post-war world (Bush 2004: ch. 2). Chiang, Churchill, and Roosevelt discussed Taiwan's status when they met in Cairo in November of 1943. The statement released to the press reported a decision that "all the territories Japan has stolen from the Chinese, such as Manchuria, Formosa, and the Pescadores, shall be restored to the Republic of China" (*Cairo Communiqué* 1943).

The "Cairo Declaration" was little more than a press release summarizing discussions. However, it became the basis for a series of documents that served to make the cession of Taiwan to China look increasingly official. When the Potsdam Declaration was issued in July of 1945, it charged that "The terms of the Cairo Declaration...be carried out" (*Potsdam Declaration* 1945). In the surrender document signed a month later, Japan agreed "to carry out the provisions of the Potsdam [and by extension Cairo] Declaration in good faith" (National Archives & Records Administration 1945).

Acting on behalf of the Allies and under orders from Supreme Allied Commander Douglas MacArthur, ROC troops accepted the Japanese surrender on Taiwan on October 25, 1945 (Romberg 2003: 2). As far as the KMT was concerned, this act legally returned the island to China, and, from all appearances, the United States accepted this claim. The small American consulate in Taipei was instructed to report to the US embassy on the mainland, since Taiwan was now China (Kerr 1965: 345). Yet the wartime declarations relied on two premises: that the Allies would remain united and that the government of post-war China would be acceptable to all of them. Over the next two years, these assumptions evaporated, and Taiwan's status became uncertain. Since the United States had taken the lead in ceding the island to KMT-run China, American policy led the international response to the Taiwan issue.

By 1947, the Truman administration was thoroughly disillusioned with the situation in China. The KMT was losing its mandate to rule. As the United States stepped back from the developing civil war with the communists, it also modified its position on Taiwan. "The transfer of sovereignty over Formosa to China 'has not yet been formalized,'" said Acting Secretary Dean Acheson in one early telling comment to a US senator (Risenhoover 2007). For the next two years, policy toward the ROC government and the island it occupied was hotly debated in Washington. Acheson was anxious to cut American ties with Chiang Kai-shek's government. He sought to keep his options open in dealing with a new communist government in China and to minimize any backlash on the mainland due to continued American support for a failed government.

The military (with the important exception of Douglas MacArthur in Tokyo) had little patience for Chiang but did value the geographic potential of the island. However, because of post-war demobilization, the Pentagon was not ready to take on responsibility for it. As the Nationalists retreated to Taiwan over the course of 1949, Washington explored the possibility of alternative ROC leadership and even Taiwanese independence, but concluded that there was no alternative to Chiang. However, this did not mean that the United States would continue to support the ROC government; in October 1949, Washington announced that it would not contribute to Taiwan's defense and that any further American aid depended on political reform (Accinelli 1996: chs 1 and 2).

On January 5, 1950, President Truman seemingly closed the book on Taiwan, announcing that the US would cease "aid or advice" to the ROC, and supported the Cairo Declaration returning Taiwan to mainland China. The president added:

> The United States has no predatory designs on Formosa, or on any other Chinese territory. The United States has no desire to obtain special rights or privileges, or to establish military bases on Formosa at this time. Nor does it have any intention of utilizing its Armed Forces to interfere in the present situation. The United States Government will not pursue a course which will lead to involvement in the civil conflict in China. (Truman 1950a)

With intelligence estimates predicting the fall of the island to communist forces by the end of the year, Truman's statement implied that the United States was ready to see the island brought under the new mainland government. However, during the spring of 1950, the administration's sangfroid began to erode as Taiwan's strategic importance increased. China was moving closer to the Soviet Union and becoming more hostile to American interests in Asia. On the eve of the invasion of South Korea by North Korea in June of that year, proposals for reconsidering American policy were circulating (Garver 1997: 24–31).

When the Korean War broke out in June of 1950, President Truman revised the official American position on Taiwan and declared "the occupation of Formosa by Communist forces would be a direct threat to the security of the Pacific area." He ordered the Seventh Fleet to prevent an attack on the island and called upon the ROC "to cease all air and sea operations against the mainland." Qualifying the wartime declarations, he announced that "the determination of the future status of Formosa must await the restoration of security in the Pacific, a peace settlement with Japan, or consideration by the United Nations" (Truman 1950b).

This was a dramatic policy reversal: the United States was both intervening in the Chinese Civil War and declaring that the return of Taiwan to the mainland would not be automatic. However, Truman's statement made reference only to the *island* of Taiwan and not to the ROC government that occupied it (Bush 2004: 86–9, draws this distinction). Washington was denying Taiwan to the communist mainland while simultaneously refusing to support the island's Nationalist regime. Washington seemed to recognize that neither the ROC nor the newly established People's Republic of China possessed the island. Rather it was held that the KMT government was there at the request of the Allies to accept the Japanese surrender and that, like all territories taken from Japan, "its legal status cannot be fixed until there is international action to determine its future" (Risenhoover 2007).

However, this approach did not last long. Over the course of the next two years, as the Cold War in Asia developed, the United States' economic and military relationship with the ROC gradually revived, and Washington recognized it as the legitimate government of China. Yet this did not mean that title to the island officially passed to its KMT rulers. Taiwan's status remained unsettled despite the signing of the San Francisco Peace Treaty between the Allies and Japan in 1951, and despite the separate Taipei Treaty with the Japanese the following year. There was no Chinese representation at the former conference due to disagreement over whether the PRC or the ROC would represent the nation. In the resulting treaty, Japan renounced "all right, title and claim to Formosa and the Pescadores," without naming a recipient of the forfeited claim (Taiwan Documents Project 1952). The latter peace agreement – a bilateral treaty between the ROC and Japan – simply reiterated the language of the previous year's treaty regarding Taiwan. More importantly, despite its recognition of the ROC as the lawful government of China, the United States still did not acknowledge the ROC's legitimate possession of the small island from which it claimed to rule all of China. In Washington's view, Taiwan's status remained undetermined.

Within a year of its establishment in 1949, the People's Republic of China began to push its claim to the island. In the fall of 1950, as Chinese and American troops fought in Korea, PRC representative Wu Xiuquan appeared before the United Nations to complain regarding "aggression against China's Taiwan by the United States government" (Chiu 1973: 231). He charged that Washington intended to prevent the island's "liberation" by the interdiction of its navy in a "full-scale, open invasion of Taiwan" by the American military. In his presentation to the Security Council, Wu outlined the PRC claim to Taiwan that remains essentially unchanged to this day. He contended that the island had long been an "inseparable part" of the territory of China that had suffered fifty years of Japanese rule. This fact, Wu argued, had been acknowledged by wartime documents (the Cairo Communiqué, the Potsdam Declaration, the Japanese surrender document) and confirmed by President Truman in his statement of January 5, 1950. The PRC, as the ROC's mainland successor, thus claimed to be the "China" which was entitled to the post-war return of Taiwan. Taiwan's "liberation...[was] entirely China's domestic affair." The

American neutralization of the island was, thus, a "criminal act of direct armed aggressive war" on China's sovereign territory (ibid.: 232).

CONCLUSION: THE SIGNIFICANCE OF HISTORY

By the beginning of the 1950s, the elements that would create the conflicted triangular relationship of the future were in place. It was in these years that two crucial issues emerged: Taiwan's legal relationship with the mainland and the nature of the island's ruling government.

On the status of the island, the ROC and the PRC ironically agreed that it was an integral part of China while differing, of course, as to which government ruled the combined island and mainland. Both affirmed the validity of the wartime pledges. As the victor in the civil war, the PRC saw itself as the successor government of China. The restoration of Taiwan to China thus became the essential unfinished task of the communist revolution. The ROC, which claimed still to be the sovereign government of *all* China, saw itself as the rightful heir to wartime agreements. Its unfinished task was to retake the mainland, using the island as a base.

The American stance was paradoxical and differed from the positions taken by the two would-be rulers. The US recognized the ROC as the sovereign government of China and vigorously defended its status as such against those countries that preferred to recognize the mainland regime. However, events during the spring of 1950 had moved the US to consider the status of the KMT's island base to be "undetermined," despite earlier wartime declarations. The American position was that the ROC was the government of all of China whose capital and administrative offices were located somewhere that Washington considered not legally a part of China or of anywhere else!

The triangular politics of post-war cross-strait relations were thus fraught with complexity from the very beginning, with the principal actors holding views that not only conflicted but also lacked internal consistency and logic. Over the decades to come, policymakers in Taipei, Beijing, and Washington would have to manoeuvre around their respective nations' prior policies and commitments as well as the restraints imposed by shifting domestic environments. However, nowhere was the impact of domestic politics more dramatic than on Taiwan. Here the divisive legacy of the establishment of KMT rule after Japanese occupation would remain a dormant issue for four decades, only to burst upon a newly democratized Taiwan.

The moment that political dissent became officially permitted, the old culture clash was formalized as an ethnic struggle in which "Taiwanese" faced off against "Mainlanders." The distinctive history of Taiwan – most significantly imperial disinterest and Japanese colonization – along with persecution under the KMT regime ("common suffering") became the material from which an opposition party would seek to craft a separate Taiwanese identity. Cross-strait relations were thus complicated still further, as Taiwan's citizenry explored a cultural identity as ambiguous as their island's political identity.

2 | The Cold War in Asia and After

As we have seen, the end of the war in the Pacific did not settle the status of Japan's former colony of Taiwan. Despite wartime plans and a final peace treaty with Tokyo, unresolved differences regarding its relationship with China and its place in the international political system remained. They would soon become further complicated by enmeshment in the developing Cold War in Asia.

THE TAIWAN STRAIT: HOT AND COLD WAR, 1950–1971[1]

During the two decades preceding the dramatic changes that came with the diplomacy of the Nixon administration in 1971–2, the stance of the Chinese leadership changed little from that taken by Wu Xiuquan at the United Nations. In the words of Zhou Enlai, speaking to the first National People's Congress in 1954, Taiwan was "China's sacred and inviolable territory and ... no United States infringement or occupation will be tolerated," and the goal was to "strive to liberate Taiwan and eliminate the traitorous Chiang Kai-shek gang" (Chiu 1973: 249).

Until the spring of 1955, this was the dominant tone in mainland statements. The emphasis was on "liberation," meaning the use of force to solve a purely domestic problem. That spring, at the same time that Zhou Enlai attended the non-aligned summit at Bandung and offered to discuss the tensions in the area with the United States (see below), he offered "to negotiate the status of Taiwan with the 'Taiwan authorities.'" A year later, Zhou announced "on behalf of the government" a willingness "to negotiate with the Taiwan authorities on specific steps and terms for peaceful liberation." Similar messages, some with greater specificity regarding the terms for Taiwan's incorporation, were also passed on to the KMT through third parties claiming to speak for Beijing.

On Taiwan, the president, Chiang Kai-shek, acted as if history had stopped in the late 1940s. The KMT, he claimed, was preparing for a "fourth military campaign" to finally defeat the communists. The ROC denied that it was an "exile government"; rather, it was the government of the whole nation, large parts of which were temporarily occupied by "rebels." Chiang's goal was a return to the mainland. His basic assumption, like that of the mainland leaders, was that the cross-strait conflict was a domestic affair to be resolved by military means – either in the form of an invasion from Taiwan or as part of a broader confrontation with global communism. This posture forbade any interaction with the mainland. Trade was impossible and negotiations on any subject whatsoever were off-limits – "the loyal and the traitorous cannot coexist," Chiang insisted. And so, almost immediately after settling into Taiwan, the KMT began to plan for the day when a full-scale invasion would herald its return to the mainland it claimed to rule.

The outbreak of the Korean War and the Chinese intervention had prompted a reluctant United States to re-establish its relationship with a KMT government that it had earlier considered politically and militarily bankrupt. The relationship developed during the 1950s and eventually was formalized in the Mutual Defense Treaty of 1954. American policy established three objectives in the Taiwan Strait area: to maintain a peaceful, stable environment, to assure the survival of the ROC government on Taiwan as a symbol of democracy, and to demonstrate American determination to support its allies in Asia. Besides the umbrella of the defense relationship formalized in the 1954 treaty, the alliance with the United States brought essential economic benefits and international recognition to the ROC. American aid played a crucial role in the reconstruction of Taiwan. Moreover, after the PRC joined the Korean War, and despite the objections from many of its allies, Washington's steadfast opposition to Beijing's entry into the United Nations both preserved the ROC's international status and gave support to its claim of representing all of China.

However, for the United States, a major concern was that Chiang's driving ambition to regain the mainland would undermine the fragile peace in the Taiwan Strait and entrap the United States in an unwanted conflict with the PRC. Thus, the American support for Taiwan was intended to serve a second and somewhat contradictory purpose – to use the influence gained from the alignment, and later alliance, for the dual purpose of frustrating Chiang's ambitions and deterring the mainland.

These objectives were difficult to reconcile. The concern regarding entrapment in an ROC-provoked conflict suggested that US alliance objectives centered on a narrowly defined commitment to the defense of the island. However, there was a countervailing concern that this might erode ROC morale or tempt the mainland to act. This contributed to more aggressive posturing which raised the chance of exactly the entrapment that the United States sought to avoid. This paradox was evident during two mainland–Taiwan confrontations that occurred in the decade after the communist victory, when protective American policies nearly resulted in entrapment in a cross-strait military conflict.

CRISIS IN THE TAIWAN STRAIT

These confrontations were centered around the "offshore islands" – three island groups, Jinmen (Quemoy), Mazu (Matsu), and Dachen, close to the mainland (Jinmen within sight) but more than 100 miles from Taiwan. Their proximity to the mainland made them more vulnerable to PRC attacks than Taiwan.

In September of 1954 and August of 1958 the mainland began to shell the islands, precipitating cross-strait confrontations that lasted until early 1955 in the former case and until the end of the year in the latter. Although the PRC's motives for initiating the confrontations with the KMT differed in each case, in both the use of force was intended more to influence or test American policy in the area than it was to resolve the conflict with the KMT. Indeed, rather than changing the status quo in the strait (except for the KMT's loss of the Dachen group), these crises twice threatened to spark a Sino-American conflict, with Washington rattling nuclear weapons.

During each crisis, Washington considered the islands to be a strategic liability for the defense of Taiwan. The Mutual Defense Treaty of 1954 had been intentionally limited to the main islands of Taiwan and the Pescadores. However, since the treaty was completed during the first crisis, President Eisenhower eventually sought, and received, a congressional resolution in February 1955 that allowed American aid in defense of the islands if an attack appeared to be the precedent to an assault on the main islands. It was a strategy that was intended to keep all options open while at the same time leaving the mainland and Chiang "guessing" as to Washington's response. Still, in both crises, because of the risks of a confrontation with Beijing that could result from the defense of these islands, Washington pressed Chiang to withdraw the exposed forces there.

Besides improving the security situation for Taiwan and lessening the chances of conflict, Secretary of State John Foster Dulles had another motive in pressuring Chiang to pull back to Taiwan. By increasing the geographical distance between the ROC and the mainland, he hoped to defuse the situation in the strait by securing the division of China similar to that which had developed in post-war Korea and

Germany. However, Chiang refused the demands of the Eisenhower administration. The islands were a forward base for the defense of Taiwan as well as a site from which to raid the mainland and carry out propaganda activities. It also represented a commitment to retake the mainland. During both crises Chiang insisted that any retreat from the islands would fatally undermine morale on Taiwan.

The Eisenhower administration was left with little choice. If the existence of Taiwan was essential to US objectives in Asia, and if this required maintaining morale on the island – which, in turn, depended upon the continued hope of regaining the mainland – then American policy would have to give some semblance of support to such aspirations, even if that might lead to American involvement in the conflict. In the first crisis of 1954–5, the Eisenhower administration conceded in private to both Congress and the ROC its willingness to defend Jinmen and Mazu and spoke openly of the possible use of nuclear weapons on what were referred to as Chinese "military targets" (Chang 1990: ch. 4).

In 1958, once again, the dual imperatives of maintaining morale on Taiwan and American standing in Asia led the Eisenhower administration to include the defense of the offshore islands and the use of nuclear weapons in its deliberations. Indeed, this time Washington went as far as to aid in the supplying of the island. Yet, once again, despite initial mainland calls to "liberate Taiwan," the confrontation never expanded beyond the seas and skies offshore, and by the end of the year the immediate crises had passed.

And so, in less than a decade, American policy in the Taiwan Strait had undergone a total reversal. It had gone from abandonment of the KMT regime and considerations of how to detach the island from the control of that regime to a mutual defense treaty with the ROC and two confrontations with the mainland, intended to maintain KMT control of an island the status of which Washington still considered to be "undetermined."

It would be more than three decades before regular contacts between Taiwan and the mainland would take place. In the intervening period, cross-strait relations would remain tense but relatively peaceful, with little contact of significance between the former enemies in China's Civil War. On the other hand, until the 1990s, to the extent that there were relations, they were between the United States and the People's Republic of China, as both sides in this leg of the triangle sought to adjust their relationship and avoid a dangerous conflict.

AFTER THE CRISES: FRUITLESS TALKS AND IMPORTANT READJUSTMENTS

The most dramatic indicator of these winds of change in the strategic triangle in the Taiwan Strait consisted of the Sino-American talks that began in April 1955. Between 1955 and 1970, 136 meetings were held at the ambassadorial level, first in Geneva, Switzerland, and then in Warsaw, Poland.[2] Although the ostensible reason for the talks was the release of Americans held in China, Secretary of State John Foster Dulles also sought to defuse the troublesome situation in the Taiwan Strait. While the initial meetings were concerned with repatriation issues, by October 1955 talks had passed on to a discussion of the "Taiwan area." In the end, they resolved nothing. However, they did serve to identify many of the irreconcilable differences between the two sides that would persist well into the twenty-first century.

The United States' objective was to produce an agreement with parallel declarations in which both sides, "without prejudice to the pursuit of each side of its policies by peaceful means," agreed that, "[i]n general and with particular reference to the Taiwan area[,] ... [each side] renounces the use of force, except in individual and collective self-defense." Should the Chinese accept, Dulles would, in effect, have achieved a cease-fire in the area with no change in the US treaty relationship with Taiwan.

This was, not surprisingly, unacceptable to the Chinese. They offered to hold talks regarding the "tense situation" in the Taiwan area as a whole, but only if such negotiations were held at the level of foreign ministers. If not, the Chinese side would discuss only the peaceful settling of Sino-American "disputes." What was absent from Beijing's proposals was any clear willingness to include such matters as a renunciation of force, the American right of self-defense in reference to the "Taiwan area," or any other issue regarding policies toward Taiwan. The PRC would not make any pledges about how this "domestic" matter might be settled.

Specifically, the Chinese held that the ROC was not an international actor. It could not conclude a treaty. The Mutual Defense Treaty, thus, was invalid, and the presence of US forces in the area constituted an act of aggression and occupation committed against the Chinese homeland. Moreover, any discussion of renunciation of force, the Chinese contended, would be to freeze an unacceptable status quo and continue China's humiliation by imperialist forces. By 1957, the incompatibility of the positions taken by each side became clear, as did the realization that there would be no agreement to ameliorate the dangerous situation in the cross-strait area. This was confirmed when the two sides met during the 1958 crisis. At this time, the talks in Warsaw simply reprised the empty exchanges of the early discussions.

In 1962, the Taiwan issue took center stage during the ambassadorial meetings once more. However, this time, rather than demonstrating the irreconcilable differences between the two sides, the meeting presaged a future readjustment in American policy. In that year, during a period of domestic crisis caused by the Great Leap Forward and an international dispute with India, the Chinese became concerned over a Taiwanese military build-up and asked to meet with the American ambassador in Poland, John Moors Cabot. Cabot informed the Chinese ambassador "that the US government had no intention of supporting any GRC [Government of the Republic of China] on the mainland under existing circumstances...[and that] the GRC committed not to attack without our consent." There was no attack from Taiwan. However, Cabot's response and other gestures by the United States were reflective of a growing sentiment in the administration of John F. Kennedy that it was time for the United States to extricate itself from the Chinese Civil War (Goldstein 2001: 228).

However, it was only during the Johnson administration that efforts to improve relations with the mainland were seriously considered, and, ironically, they were tied to the expansion of the war in Vietnam. Throughout the period of American escalation, one of the Johnson administration's major preoccupations was the possibility of Chinese intervention in that war. Restraining Taipei to avoid entrapment remained at the core of American alliance objectives, but there was also consideration of conciliatory gestures to China. However, by that time the Sino-American dialogue over Taiwan was no longer possible. The Chinese Cultural Revolution was beginning. Aside from occasional blustering about the liberation of Taiwan, there was little movement in mainland policy regarding the triangle in the Taiwan Strait.

This was not the case with the ROC. With hopes of mainland recovery dashed by Washington's uncompromising determination to restrain its ally, there were tentative signs of new thinking in Taipei. By the early 1960s, the ROC had begun the process of economic trans-formation that would eventually make it one of the "miracles" of Asian industrial development. Promoting economic development and secur-ing ties with the rest of the world under the umbrella of the security alliance with the United States was a new objective of the ROC. This reorientation toward economic development on the island had obvious domestic political implications. The KMT government's earlier reluc-tance to devote resources to one province's development sprang from its refusal to acknowledge Taiwan as anything more than a temporary way station for the party's return to the mainland. The new commitment to develop the island suggested a more permanent residence.

There were also signs that, toward the end of his life, Chiang Kaishek was rethinking his ambition to return to the mainland. By early 1967, the United States embassy was reporting that the mood in Taiwan was turning away from mainland recovery, and Chiang was calling for greater use of political means to effect change on the mainland (Goldstein 1999: 62).

After Richard M. Nixon assumed the presidency in 1969, the relationship between China and the United States entered a new stage. However, it should be no surprise that Taiwan would prove to be not only the most prominent item on the agenda of discussions but also *the* major obstacle to the development of the relationship thereafter. Sino-American relations remained focused on the Taiwan Strait triangle.

RAPPROCHEMENT WITH BEIJING AND THE INTRACTABLE TAIWAN ISSUE[3]

During the Nixon presidency, the Vietnam War, policy toward the PRC, and the Taiwan alliance were interwoven. Determined to extricate the United States from the war in Vietnam, the administration at first sought improved relations with Beijing, not simply to put pressure on North Vietnam at the negotiating table but also to ensure a more stable Asia after the United States withdrew. Over time, Washington's impetus for an improved relationship – and the basis for mutual reconciliation – also came from shared concerns over Soviet intentions and the belief that rapprochement with the PRC would provide the United States with additional leverage in relations with Moscow.

The result of this reorientation was that the mutual defense treaty with Taiwan lost much of its rationale. Beijing was no longer the adversary. Closer relations with the mainland had become a keystone of

American foreign policy. During the Nixon and Ford administrations (1969–77), the United States was negotiating the terms of a strategic realignment with the former object of the mutual defense treaty with Taiwan. However, to gain that realignment, the United States would have to accept the terms laid down at the ambassadorial talks: the end of the mutual defense treaty and no commitment that China would abjure the use of force to secure the island's reunification. In short, the Nixon and Ford administrations would have to move on a course that could logically lead only toward abandonment of the ROC and extrication from the Taiwan Strait triangle.

However, that was simply not an option for either administration. Although domestic public opinion favored improved relations with China, there was opposition within Congress and the Republican Party to ending the commitment to the defense of Taiwan. Finally, as the American withdrawal and, ultimately, the fall of South Vietnam unfolded between 1973 and 1975, there was considerable opposition to the idea of abandoning yet another ally.

Given the shared interest that China and the United States had in regard to the Soviet Union, potential turmoil in East Asia after the Vietnam War, and the Pakistan–India conflict, it is not surprising that the basis existed for consultations on international questions. Achieving such consultations alone would have been a dramatic development and would certainly have added luster to Nixon's foreign policy record at little domestic or international political cost. It would appear that this was the intent of the administration when it began the process that would eventually result in the secret diplomacy of Henry Kissinger in July 1971 and the very public visit by the president six months later.

Indeed, there was relief in Washington when, during the exchange of secret messages before Kissinger's trip, the Chinese appeared to be focused on the withdrawal of American forces in the area. This was seen as an issue that could easily be resolved without touching upon the more fundamental – and difficult – issue of Taiwan's status and Washington's overall relationship with the ROC. Most American forces on Taiwan were supporting the effort in Vietnam and were scheduled to be withdrawn when that conflict ended. Moreover, withdrawal was consistent with the Nixon Doctrine regarding self-help in Asia. At least one administration study suggested that, unless Taiwan was included, it might appear to Beijing that the United States intended to separate it from China (Romberg 2003: 23).

The briefing book prepared for Kissinger's trip in July 1971 reflected the expectation that differences regarding the political aspects of the Taiwan issue would not be a major obstacle to tackling other global topics. Moreover, it was apparent that President Nixon did not want any discussion of the overall American relationship with Taiwan. On the cover of the briefing document he scribbled, "Don't be so forthcoming on Taiwan – until necessary," and in his conversation with Kissinger on the eve of the latter's departure he repeated this and "asked him to review the entire discussion of the Taiwan issue so that we would not appear to be dumping on our friends and so that we would be somewhat more mysterious about our overall willingness to make concessions in this area" (Fatal Politics 1971; US Department of State 2006: doc. 137).

However, at the very beginning of the secret talks, his interlocutor, Premier Zhou Enlai, raised the stakes. He immediately dangled before Kissinger the prospect of diplomatic ties while making it clear that issues related to the status of Taiwan and American relations with the island were the major obstacles. This demarche was unexpected by the American side. In making this proposal, Zhou was complying with a Politburo document which made it clear that the issues Beijing was prepared to raise with the Americans went well beyond the withdrawal of troops (although this was considered to be the minimum requirement for a presidential visit). According to this secret document, "diplomatic relations" were not to be established until this demand was met *as well as* recognition that the Taiwan issue was an internal affair; a severing of relations with the ROC; recognition of the PRC as the sole legal government of China; and an abandonment of any attempt to pursue a "two Chinas" or "one China and one Taiwan policy." In contrast to Washington's expectations, the Chinese intended to raise the full spectrum of post-war Sino-American differences over Taiwan (Romberg 2003: 26–7).

Contrary to the claim in Kissinger's memoirs that Taiwan "was mentioned only briefly in the first session" of his meetings with Zhou, the transcript of the talks reveals that, apparently to his surprise, the topic of Taiwan came up almost immediately and dominated the early discussion. At the start of the meeting, Zhou invited Kissinger to present the topics for discussion. Although Taiwan was the first topic he mentioned, Kissinger clearly sought to keep it restricted to the relatively easy (for the United States) issue of troop withdrawal. He acknowledged that Taiwan was the Chinese side's "principal concern" and added, "Mr Premier, you have defined this as the withdrawal of U.S. forces from the Taiwan Straits [*sic*]" (Romberg 2003: 26–7). Zhou clearly would not accept this narrow focus on the Taiwan issue. His response established that the central issue for the Chinese side was the American failure to treat China as an equal. Washington had refused to acknowledge Taiwan as a part of China and had signed an "illegal" treaty with the regime there. He demanded: "in recognizing China, the U.S. must do so unreservedly. It must recognize the PRC as the sole legitimate government of China and not make any exceptions" (ibid.: 29).

In the meetings that followed, Zhou pressed for movement toward "normal relations" (taken as diplomatic relations) and was uncompromising in his demand that the fundamental prerequisite would be both the severing of relations with the government on Taiwan and the recognition that the island was an "inalienable part of China." Moreover, he demanded a statement that the United States "does not support a two Chinas or one China, one Taiwan policy and does not support the so-called Taiwan Independence Movement" (US Department of State 2006: doc.140). It was, in effect, a demand for the end of the American relationship with Taiwan.

From the transcript of the talks, it appears that Kissinger was genuinely taken aback by the direction the Taiwan topic had taken. He noted that Zhou's presentation had gone "beyond some of the communications we have previously exchanged" and that he had now "spoken of certain official political declarations" (US Department of State 2006: doc. 139). Zhou was introducing topics that the president had wanted to avoid and which Kissinger, apparently, was unready to address. However, the Chinese side was insistent. Kissinger had to respond, and, in doing so, he exceeded his mandate. He acknowledged that four of the five requirements mentioned by Zhou Enlai could be accomplished by President Nixon "within the near future": first, no support for the Taiwan Independence Movement; second, no more official statements that the status of Taiwan was undetermined; and, third, no support for two Chinas or one China, one Taiwan. Kissinger assured Zhou that the fourth requirement, recognition by the United States that "Taiwan belongs to China," would "take care of itself as a result of the other three points." The fifth, recognition of the PRC as the "sole legitimate government of China," would have to wait "until after the elections" in the fall of 1972 (ibid.).

In sum, in the initial conversations with Zhou Enlai, Kissinger appears not to have been very "mysterious" about the Taiwan issue. By the time he left, he had set a target date for formal diplomatic relations, all but conceded the issue of Taiwan's status as a part of China, and suggested that Washington would not stand in the way should developments lead to union with the mainland. As he emphasized in his conversations with Zhou, the basic "direction" of relations had been set; the only issue was one of "timing." With apparent confidence, Kissinger assured him that "We can *certainly* settle the political question within the earlier part of the President's second term. *Certainly* we can begin evolution in that direction before" (US Department of State 2006: doc. 139; emphasis added).

In subsequent meetings before Nixon's visit, American apparent acceptance of Chinese demands continued (Tucker 2009: 244). In October 1971, Kissinger went further in meeting the Chinese position when he said that the administration's position was to "encourage" a "peaceful resolution within the framework of one China." Moreover, in response to Zhou's demand that the Mutual Defense Treaty be abrogated, Kissinger, while avoiding the question of how the mutual defense treaty might be handled before such a resolution, replied that, "[w]hen China and Taiwan become one, we can abrogate that treaty" (US Department of State 2006: doc. 162).

However, the most accommodating gesture came during President Nixon's formal meeting with Premier Zhou Enlai. The president made a concession to the Chinese position that has never been repeated since by any president or major American official. In his first substantive meeting with Zhou, the president immediately presented "[p]rinciple one. There is one China and Taiwan is a part of China. There will be no more statements made – if I can control the bureaucracy – to the effect that the status of Taiwan is undetermined" (US Department of State 2006: doc. 196). In short, within the space of less than two years, the Nixon administration had given not only pledges that must have appeared to the Chinese as a commitment for eventual acceptance of their conditions for establishing relations but also a timetable for when that would happen.

With President Nixon's re-election in the fall of 1972, the United States appeared ready to move ahead on the road to recognition. In a letter to Zhou Enlai on January 3, 1973, the president wrote that he wished "to reaffirm our intention to move energetically toward normalization of our relations. Everything that has been previously said on this topic is hereby reaffirmed" (US Department of State 2006: doc.1).

In his meetings with Zhou Enlai a month later, Kissinger spoke of a "substantial reduction" of military forces on Taiwan and noted that Washington was "going very slow about giving new military equipment" to Taipei. He noted, "after 1974 [mid-term elections] we want to work toward full normalization and full diplomatic relations with the People's Republic of China." Kissinger did repeat the importance of an "understanding that the final solution would be a peaceful one" and the fact that the United States would like to "keep some form of representation" on Taiwan. Zhou did not respond to these points (US Department of State 2006: doc. 18).

When Kissinger returned to Washington he was ebullient. In his report to the president, he characterized the talks as "the freest and most candid" while his "reception [was] the most cordial." He noted that, while the Shanghai Communiqué had "finessed the Taiwan problem through mutual and ambiguous compromise," something more had been accomplished during his February trip. The two sides, he believed, had "reached a clear modus vivendi – on our part, *continued concrete evolution toward full relations with all its implications*; and on their part *patience and pragmatism*." In this vein, the Chinese agreement during that trip to establish a liaison office was seen by Kissinger as "bending the sacred 'one China policy' in order to ease our predicament" (US Department of State 2006: doc. 18; emphasis added).

What was the American predicament? In his discussions with Zhou, and especially in the negotiations over the joint communiqué, Nixon stressed that he took the danger of a political backlash very seriously, warning that politics at home presented not only a threat to his "political survival" but also "a danger to the whole initiative." However, to keep the momentum of rapprochement going, the president and Kissinger were making commitments in private that would be threatening in both respects if made public, while assuring the Chinese that they could do more than they could say openly. In other words, the strategy that developed from 1971 to 1973 was to *say* in secret talks much more than it was possible to *do* and, by doing so, hopefully keep the Chinese side engaged. The most blatant example of this is the contrast between the very oblique American statement in the public Shanghai Communiqué that "all Chinese on either side of the Taiwan Strait maintain there is but one China and that Taiwan is a part of China. The United States Government does not challenge that position," and the unambiguous statement by President Nixon in private that Taiwan was a part of China.

Moreover, in that specific regard as well as more generally, the statements made in public about the meetings in no way conveyed the extent of the apparent concessions being made by Kissinger and Nixon in private. Quite to the contrary, throughout the secret negotiations they spoke with confidence, both in public and to Taiwan's representatives, about the future of Taiwan, even providing assurances that the United States would not abandon an "old friend." One must marvel at the hubris reflected by the apparent confidence on the part of the president and his national security adviser that the relationship with Taiwan as well as the political clout of its supporters in the United States could be overcome in the last four years of the administration.

One clue to how Kissinger considered this might be accomplished is found in one of his memos to the president, where he wrote:

> Taiwan is a problem we should be able to control, both internationally and domestically, as we continue to add to the handwriting on the wall and condition our audience. However, we should be under no illusions that our final step will be anything but painful – there are few friends as decent as our allies in Taiwan. (US Department of State 2006: doc. 18)

Given the treatment of Taiwan during the rapprochement with China, it is hard to see the final statement about Taiwan as much more than crocodile tears perhaps written for the record. Moreover, the comment about the "handwriting on the wall" seemed to reflect Kissinger's belief that, as the United States moved closer to relations with China, the government on Taiwan would conclude that it had no option but to accommodate Beijing, thus solving Washington's predicament.

However, the future would show this strategy was not only flawed but involved costs in several respects. Chinese interlocutors had heard pledges that met their long-standing demands. The result was heightened expectations that these demands would be met, only to be followed by disappointment and impatience when they were not. In addition, the strategy would only complicate the process of addressing the challenge of domestic politics. By making public statements that minimized the concessions being offered in the talks with the Chinese, the administration was neglecting the kind of domestic political groundwork that would be necessary to achieve the reversal anticipated.

In any event, after February of 1973, preoccupation with Watergate made it impossible for the administration to fulfill its pledges. As it turned out, President Nixon's successor, Gerald Ford, an unelected vice-president who took office as president following Nixon's resignation in August 1974, faced political challenges from the conservative wing of the Republican Party and public doubts regarding America's global power following the end of the Vietnam War, and could not risk moving the relationship with China forward. Although he managed a trip to China in 1975, there was little of substance discussed and even less accomplished. For the next five years, until the Carter administration's recognition in 1979, Sino-American relations progressively worsened, as it seemed to the Chinese that, rather than moving forward on Kissinger's pledges, the administration was, at best, stalling and, at worst, backing away.

For Taiwan, Washington's China initiative was the final nail in the coffin of its aspirations to regain the mainland and a blow to the ROC's international position. Although the US–ROC military alliance would remain intact until 1979, its significance as a symbol of American military support was rapidly vanishing. Now it was important as a defensive shield behind which the island could develop its economy and nothing more. The American initiative also affected Taiwan's inflated international standing. Washington's increasingly intimate dealings with Beijing released other countries from the obligation to support the post-war fiction. The 1971 admission of China into the United Nations was the most immediate sign of this. Taiwan's bilateral relations also suffered. In 1970, fifty-three nations had recognized the PRC and sixty-eight the ROC. By 1977, the figures were 111 and twenty-three, respectively.

The KMT's policy response to these changes was a limited reconsideration of its mainland policy. While the island was still referred to as a "bastion" for national recovery, KMT policy statements throughout the 1970s suggested that Taiwan would become a "model province," and that would appeal to the mainland population by presenting a better life that might be possible under KMT governance. This shift toward peaceful reunification and the concept of a "model province" carried with it an increased focus on building political legitimation via Taiwan's political and economic development rather than by an unlikely return to the mainland.

In 1972, Chiang Kai-shek's son, Chiang Ching-kuo, became premier and began a series of reforms. With economic policy and political institutions no longer oriented toward sustaining the island as a military jumping-off point, greater attention was now paid to the economic welfare and political rights of the local population. Chiang initiated small steps toward political reforms, such as local elections and by-elections for additional representatives to national bodies and the appointment of native Taiwanese to official posts. Economically, a series of major infrastructure projects were begun, and Taiwan's economy began to move beyond the earlier export of labor-intensive goods toward more sophisticated products. In the decade ending in 1978, the average per capita income grew fivefold while foreign exchange reserves grew by four times.

In response to the erosion of official diplomatic ties, the ROC adopted a policy of "all-out diplomacy" or a "pragmatic foreign policy" for the purpose of establishing "unofficial" contacts with foreign states that had recognized Beijing. Such ties involved establishing semi-governmental institutions abroad for the purpose of maintaining cultural or economic contacts and forming organizations to promote people-to-people diplomacy. In line with this policy, Taipei claimed to be maintaining relations with 114 countries and to have attended 258 international conferences by 1974 (Chang 2005: 243–4).

CROSS-STRAIT POLITICS WITHOUT CROSS-STRAIT RELATIONS

Despite the signs of adjustment and innovation in Taiwan's domestic and foreign policies, and despite the dramatic change in Sino-American relations, the ROC's mainland policy showed no signs of change. The issues of the relationship between Taiwan and the mainland as well as the ROC's place in the international system were still being thrashed out, not across the Taiwan Strait but across the Pacific, as the United States and China moved from confrontation to negotiation in the area.

However, despite the incentives for both sides to move the bilateral relationship forward, mutually incompatible positions on these two issues stymied real progress. These years of dramatic development in Sino-American relations had also demonstrated how little flexibility each side had in its approach to Taiwan. In the next decade, this incompatibility would remain despite the achievement of Sino-American normalization.

3 | Normalization and New Problems

Nixon's dramatic China initiative had done little to solve the Sino-American deadlock over Taiwan. The substance of the differences had not changed since the 1950s and 1960s, and the various discussions had done little to resolve them. In fact, it could be argued that ground was actually lost, as the Chinese became more suspicious of American sincerity.

In 1979, normalization of Sino-American diplomatic relations was achieved. The United States finally ended its alliance with the ROC on Taiwan and formally recognized the PRC as the legitimate government of China. The expectation was that this would both remove the tensions between Beijing and Washington over cross-strait relations and allow the two sides of the strait to end the civil war. However, what came to be called "normalization" established fuzzy and contested boundaries that achieved neither.

This chapter explores the origins and nature of these boundaries. In the next chapter, we will discuss how they functioned when changes in Taiwan created a new, more complex political situation on the island that challenged the shaky Sino-American modus vivendi that had emerged during normalization.

NORMALIZATION: THE CARTER ADMINISTRATION BREAKS THE DEADLOCK

Soon after Jimmy Carter became president, his assistant for national security, Zbigniew Brzezinski, argued that the experience of previous administrations had shown that improved Sino-American relations were essential for dealing with the Soviet Union and that such improvement required Washington's "flexibility" on the Taiwan issue.[1] He asked, rhetorically, "What leverage will we surrender over the Soviets should we fail to demonstrate continued movement on the Taiwan issue?" President Carter agreed, commenting that, although he would not approach the Chinese with the "almost abject" attitude of previous administrations, he would adhere to commitments made by them.

In March 1977, Brzezinski informed the president that he had found neither a "formal, secret agreement" nor "promises made by the Chinese" in the negotiating record. However, he reported that "Nixon and HAK [Henry A. Kissinger] repeated five points on several occasions," which he said "constituted a SECRET PLEDGE." The first "point" was the one made only in private by President Nixon and not reflected in the Shanghai Communiqué or any other public document: "There is one China and Taiwan is part of it. We will not assert the status of Taiwan is undetermined" (US Department of State 2006: doc. 16).[2] Other Chinese conditions were clear from the negotiating record and had been reiterated by the Chinese representative in Washington. These were to sever the diplomatic relationship with Taiwan; withdraw US troops from Taiwan; and abrogate the 1954 Mutual Defense Treaty. Compliance with these three requirements was not considered to be a problem. What *was* a problem for Carter (as it had been for his predecessors) were the political consequences of a sudden abandonment of Taiwan. And so the Carter administration formulated its own three conditions for normalization which would address the political consequences of appearing to bow to Chinese demands: assurance of a peaceful resolution of the cross-strait conflict; American retention of a full range of economic and cultural relations with Taiwan, including the sale of arms; and the establishment of some "informal" economic and cultural institutional link to replace the embassy.

Because in previous negotiations the Chinese had been willing to accept the idea of a mission similar to that established by the Japanese after recognition, this was considered to be the least controversial issue. However, there seemed to be little doubt that the Chinese would refuse to pledge the non-use of force in resolving the cross-strait differences. Beijing had repeatedly stated that such resolution was an internal affair.

Finally, the continuation of arms sales, considered an essential element in gaining domestic acceptance of recognition, was also seen as a difficult issue. As Secretary of State Cyrus Vance prepared for an exploratory meeting with the Chinese in August 1977, there was concern regarding the Chinese response to this particular issue, not only because of its obvious sensitivity but also because it had not previously been raised in negotiations. Still, President Carter instructed Vance to address all three issues, adding that, on the issue of arms sales, he was in favor of a "direct approach" of "leaving no doubt…that we intend to preserve Taiwan's access to sources of defense equipment" (US Department of State 2013a: 130–1).

However, when Vance arrived in Beijing he found that neither Foreign Minister Huang Hua nor Deng Xiaoping was prepared to negotiate on any of the Chinese conditions, and both viewed any attempt by the American side to do so as a delaying tactic. Vance's attempts to argue for some official status for a future American presence in Taiwan were rejected, as were his efforts to negotiate peaceful resolution. He *never* raised arms sales as instructed by the president. The visit neither proved a propitious beginning to the administration's China policy nor revealed any Chinese flexibility.

In May 1978, it was Zbigniew Brzezinski's turn. The principal rationale for his trip was to maintain consultations on strategic issues – i.e., opposition to the Soviet Union (US Department of State 2013a: 288–91). However, Brzezinski planned to express the administration's commitment to initiating a "secret approach" by the

American ambassador, Leonard Woodcock, to negotiate normalization. He was authorized to agree to China's three conditions but also ordered to reaffirm Washington's three conditions, with arms sales identified as "the most delicate aspect of the negotiation" (ibid.: 357–67).

In Beijing, Brzezinski accepted the three Chinese conditions. Regarding peaceful resolution of cross-strait relations, he acknowledged that this was a domestic issue for the Chinese, but said it would be helpful if they did not contradict Washington's expression of "hopes" for a peaceful resolution. In his approach to arms sales, Brzezinski's presentation was far less direct – even obscure. He noted that "the maintenance of a full range of commercial relations with Taiwan would provide the necessary flexibility during the phase of accommodation to a new reality in the course of which eventually one China will become a reality" (US Department of State 2013a: 439). What Brzezinski appeared to have expected Deng to understand was that arms sales (a "full range of commercial relations") were needed to satisfy potential critics ("provide the necessary flexibility") until Taiwan should officially became part of China. He softened the impact of this apparent qualification by denying that they were "preconditions"; rather, as he put it, they were a "context" that would make it "easier" to manage domestic opinion.

Deng Xiaoping's response was to reaffirm that the path to normalization required American acceptance of China's three conditions. American relations on a non-governmental basis (i.e., the "Japanese formula") were acceptable. As for peaceful resolution, he suggested the format of the Shanghai Communiqué – parallel statements where each side expressed its own views. Finally, there was no comment by Deng suggesting that he had understood that "full commercial relations" included arms sales. Once again, a discussion of arms sales had been avoided.

However, when Brzezinski met with Premier Hua Guofeng, the topic was broached explicitly for the first time in the context of a discussion of peaceful resolution. Hua noted:

> If we undertake the commitment that China will not liberate Taiwan by arms, then on the other hand the U.S. side is helping and arming Taiwan with its military equipment. What will be the result of these actions? I think it is still the creation of one China, one Taiwan, or two Chinas…We think if Chiang Ching-kuo of Taiwan did not get U.S. equipment and weapons there might have been a quicker and better settlement of this issue. (US Department of State 2013a: 452)

In his post-trip report, Brzezinski linked this statement with his own comments regarding the retention of a "full range of commercial relations" as a sign that the Chinese had understood his meaning and were suggesting a possible willingness to negotiate a peaceful resolution in exchange for arms sales restraint. However, when the negotiations started a month later, arms sales were still considered to be an unresolved, deal-breaking issue about which there was little understanding of the Chinese position or what if anything had been signaled from the American side.[3]

The Carter administration decided it would raise each of the three American conditions in order of anticipated difficulty, with the issue of non-governmental representation followed by a proposal for parallel statements on peaceful resolution and finishing with the question of commercial relations and arms sales. In mid-August 1978, Brzezinski reported that, overall, the initial Chinese reaction to the American conditions was "tough but not foreclosing." In mid-September, in a specific reference to arms sales, Woodcock reported that the issue was "difficult" to get past, but that "the Chinese do not seem to be slamming the normalization door in our face over this issue, even while sketching out a position that is substantially at odds with our own" (US Department of State 2013a: 563 n. 3).

However, when President Carter met with the Chinese representative in Washington, Chai Zemin, on September 15, he made the first explicit statement of the American position on post-normalization arms sales. The United States, he said, "will continue to trade with Taiwan, including the sale of some very carefully selected defensive arms[,]…in a way that carefully does not endanger the prospect of peace in the region and the situation surrounding China." Chai responded sharply that this was not in conformity with the spirit of the Shanghai Communiqué (US Department of State 2013a: 533). When the foreign minister, Huang Hua, met with Secretary Vance a month later, his position was uncompromising. Past arms sales, as well as sales after normalization, he said, violated the spirit of the Shanghai Communiqué and represented interference in China's internal affairs. Washington apparently "had not made up its mind" to normalize, and there would be no "relaxation" or "flexibility" by the "Chinese leaders" (ibid.: 553–4).

Despite these remarks, in the remaining two months until the announcement of normalization on December 15, the process dramatically accelerated. The last months' negotiations were focused on Washington's three conditions. Two of them had been extensively discussed and solutions were clearly within reach. Post-normalization American relations with Taiwan would be non-official, and public statements on the American side regarding the desire for peaceful resolution would not be contradicted by the Chinese but, rather, met by Chinese statements that relations with Taiwan were not subject to foreign interference. However, the issue of post-normalization arms sales had still not been negotiated: as we have seen, earlier in the fall, a few unilateral statements of intent had been made by each side but rejected by the other.

Between December 13 and 15 (the day normalization was to be announced) Woodcock met with Deng Xiaoping three times to discuss the final stages. It was during these last hurried meetings intended to wrap up negotiations that arms sales, largely avoided in earlier meetings, were finally discussed, but not resolved.

At the first meeting, Deng raised two issues. To send a message to the Soviets, he pressed to have the anti-hegemony clause from the Shanghai Communiqué, which had been a not-too-subtle barb aimed at the Soviets, included in the joint statement of normalization. The second set of issues related to the question of Taiwan. The Carter administration had decided to end the Mutual Defense Treaty with Taiwan by utilizing Article 10 in the treaty, which provided for a year's notice of termination. This meant that the treaty would be in effect during the first year of US–PRC relations, until December 1979. Deng argued that this made it appear as if "the treaty is abrogated in name but still exists in reality." Although he didn't state it outright, the clear impression was that this would infringe on the principle of one China. He made two requests to be passed on to President Carter: that Article 10 would not be cited publicly and that, during the period when the treaty was technically in effect, Washington would not sell weapons to Taiwan. It was in this context that arms sales were finally addressed directly at an authoritative level.

However, the record suggests a confused conversation on this crucial topic. Deng's focusing on sales under the Mutual Defense Treaty, which would remain valid until December 31, 1979, offered a compromise: China would agree to the use of the word "terminate" in reference to the treaty (thereby implicitly acknowledging Article 10) if the United States "will not sell arms to Taiwan during this period and also that you will not quote Article 10 in the statement on the Treaty" (US Department of State 2013a: 435 and 437).

Woodcock's report to Washington was ebullient: he sensed that Deng was impatient and "determined to pin down normalization at an early date." Indeed, he was so impatient that he made his comments on the basis of a spot translation rather than examining the document. Deng, Woodcock noted, was ready to go ahead with accomplishing normalization within a short time frame "on terms that could easily be interpreted in China and abroad as compromising long-held Chinese positions" (US Department of State 2013a: 638). However, it was not clear that Deng understood the American assumption that, while new arms sales would not take place during the termination year, they would continue *after* December 31, 1979, though Woodcock sensed he had implied this.

> Teng [Deng] was quite explicit in indicating that he was talking about a one year period. But *he did not explicitly confirm that we could resume arms sales once the Treaty had formally lost effect...... we cannot blythely assume that the Chinese have given us a green light for arms sales from 1980 on.* Nevertheless, this was the distinct implication of Teng's comments. (Ibid.: 638; emphasis added)

The next day Woodcock again met with Deng, and again arms sales came up. Deng was amenable to a presidential announcement declaring that, during 1979, the United States "would not sell any weapons or military equipment to Taiwan" but would follow through on those "previously committed or in the process of delivery." However, he objected to the specification of the date because "its inclusion carried an implication concerning what would happen in the years following 1979." In other words, Deng appeared to be interpreting the specification of the year 1979 as leaving open the possibility of sales *after* that date – exactly what Washington intended.

Woodcock's response was that the specification of the date was related to Deng's earlier request that arms sales should not take place *in the year* that the treaty was in force. Deng accepted this explanation. Woodcock finished his reporting cable with the comment that "[t]here is no doubt in my mind that we have clearly put on the record our position with respect to arms sales" (US Department of State 2013a: 641).

The record does not substantiate this conclusion. It is not at all clear that Deng understood the American position on arms sales "after normalization." The entire discussion of this issue, as well as the search for a solution, focused on *only* one year – 1979. Rather, the record suggests that, after two meetings, Deng apparently assumed that there would be no *new* sales during that year and none afterwards. However, continued arms sales would be an important part of the administration's defense against charges Taiwan was being abandoned. There was concern that the Chinese might be taken by surprise when this was clearly stated by the Carter administration and might respond in a manner that would poison the atmosphere of normalization. The White House insisted that Woodcock return to make the American position clear, adding that, while it did not expect the Chinese "explicitly to agree to such sales," it was seeking "forbearance" when the terms were announced (US Department of State 2013a: 647, n. 6).

The ambassador's immediate response was that, with the president's September 15 statement to Ambassador Chai and Minister Han's "emphatic objection," each side had placed its "position clearly on the record." Deng, he reiterated, understood the American position. While he expected Chinese objections to sales, Woodcock didn't expect them to act "contrary to their own interest" (US Department of State 2013a: 647). The White House insisted. The ambassador met with Deng for the third time in order to present a full and unambiguous presentation of the American position on post-normalization arms sales. Woodcock reported that Deng "emphatically stated that he could not agree," noting that acting Foreign Minister Han had stated China's views subsequent to President Carter's conversation with

Ambassador Zhai.[4] He explained his understanding of the last meeting with Woodcock.

> *Teng*: Did you notice that I raised the question yesterday that the U.S. would refrain from selling arms to Taiwan in the year 1979. I asked about the years following. Did it mean the U.S. would continue selling arms to Taiwan after the next year? Did you notice that?

> *W[oodcock]*: I thought you referred to whether or not these questions would be raised. I misunderstood.

Deng argued that the sale of arms "would amount to retaining the essence of the MDT [Mutual Defense Treaty]" even after it expired and that sales would undermine peaceful resolution. He warned that, "if Chiang Ching-kuo should lean on a certain powerful support, say the provision of arms, and refuses to talk to us about reunification," China would "be compelled to use force to solve the Taiwan issue." Woodcock was contrite, predicting that a change in American political attitudes toward the Taiwan question would solve this problem, as "over time the sentiment will be for unification." After venting, Deng began to explore ways of "managing the problem." He cautioned: "it is not good that soon after the establishment of diplomatic relations we should start quarreling…if this problem is created it will be an obstacle in the future." Deng agreed to proceed with normalization but stressed that, if the president created the impression of continued arms sales to Taiwan, the Chinese would make an appropriate public response – "a situation the Chinese would hope to avoid."

In his final report to Washington, the ambassador concluded: "we have come full circle on this issue. We cannot agree on the arms sales question but we can agree to disagree….Teng will not give us a free ride." Thus, the state of the arms sales issue became as reported by Brzezinski to Chai Zemin the morning of the announcement of normalization: "We are not trying to change your position, but we will not change our position."

At 9 p.m. on December 15, President Carter went on television to announce the establishment of diplomatic relations with China, pledging that this would "not jeopardize the well-being of the people of Taiwan" and that the American people would "maintain our current commercial, cultural, trade, and other relations with Taiwan through nongovernmental means." At the same time, Brzezinski made explicit that "relations" would include arms sales of "a defensive character" after the expiration of the treaty in 1979. As Deng had predicted, the Chinese side reacted in the form of a statement by Premier Hua Guofeng that China disagreed with this policy, considering it "inconsistent with the principles of normalization, with the goal of peaceful liberation, and with the promotion of regional peace and security."

However, the negative effect of the Chinese reaction was softened somewhat when, on January 1, 1979, a New Year's message sent by the Standing Committee of the National People's Congress to their "compatriots on Taiwan" promised to "respect the status quo" on the island and called for efforts to achieve reunification through negotiation as well as greater people-to-people contacts. When Deng came to Washington the next month, at President Carter's request, he reinforced that message (Romberg 2006: 25–9).

TAIWAN COPES, CONGRESS REBELS

As we have seen, since the end of World War II, the relationship between Taiwan and the mainland became an "issue" negotiated by the United States and the People's Republic of China, with Taiwan often in the position of a poorly informed or rarely consulted third party. However, American policymakers realized that its interests could not be ignored. Policy toward Taiwan had to balance three competing interests: satisfying the ROC and its domestic American constituency, maintaining credibility among US Asian allies, and accommodating mainland sensitivities on the Taiwan issue.

As we have seen, the Nixon and Ford administrations sought to balance these demands by their policy of secrecy. In private, they recognized Chinese claims on Taiwan and talked of the eventual incorporation of the island, while publicly misleading the ROC as well as its supporters in the United States. Of course, the effectiveness of this approach was never tested. Much to the frustration of the Chinese, after 1973 neither Republican administration was confident enough to face the reaction that they knew would surely result from finally moving ahead with an acceptance of the Chinese terms.

The Carter administration was equally aware of the adverse reaction at home and on Taiwan that would result from accepting the Chinese terms. However, its attempts to lessen the fallout failed on both fronts. In fact, recognition resulted in a new American posture in the Taiwan Strait that served to intensify rather than ease tensions in the area. In respect to Taiwan, the administration began the process of normalization by engaging in a candid dialogue in what was called a "conditioning process," intended to prepare the island for the ultimate end of diplomatic and security relations (US Department of State 2013a: 30 and 38). For example, both before and after the trip to China in August 1977 by Secretary of State Cyrus Vance, the government on Taiwan was briefed. Chiang Ching-kuo was told that the talks would concern normalization but also that they would be exploratory and cautious. Assurances were given that the United States would do nothing that would damage the security of Taiwan or make it impossible to maintain the "essence" of its relationship with Taipei (ibid.: 219–25).

In response to these frequent signals, Chiang appeared calm as the secret negotiations in Beijing proceeded the following year. He assured Washington that Taiwan would remain close to the United States and would never negotiate with the mainland (US Department of State

2013a: 224). However, the ROC was not entirely passive. It continued to strengthen its position in the United States, working with its friends in Congress and building support among the American public. Moreover, there were suspicions in the administration that, besides the intensive public diplomacy, Taiwan was engaged in covert efforts to gain information on the ongoing process as well as to slow its progress.

Despite these efforts to prepare the ROC for the inevitability of normalization, it clearly came as a shock to Taipei. The poor state of relations with Taiwan was evident when Deputy Secretary of State Warren Christopher arrived on the island at the end of December 1978 to begin charting the post-normalization relationship and was greeted by violent demonstrations. The delegation sought to ease tensions, assuring their hosts that most agreements and treaties (not the Mutual Defense Pact and related treaties) between Taiwan and the United States would remain in effect. In addition, the "people of Taiwan" would be accorded the same rights and privileges as nationals of other countries.

The proposal that proved most controversial at these meetings was that each side would "put into operation by not later than February 28, 1979, a new instrumentality created under its domestic laws which would neither have the character of, nor be considered as, official governmental organizations." The Taiwan side rejected this stipulation because it failed to recognize the sovereignty of the ROC and, by implication, the legitimacy of the KMT government. In his report to the State Department and the White House, Christopher ascribed the resistance he encountered, and the expectations that Taipei had regarding the impact that normalization might have, to domestic politics in the United States and, most of all, the "chance of gaining congressional acceptance of their position, or some variant of it" (US Department of State 2013a: 683). Their expectations proved to be not far off the mark.

From the very beginning of normalization, hardly a discussion among those managing the process in the executive branch failed to mention the importance of keeping Congress abreast of developments. It was conceded both that legislation would be necessary to implement the new relationship with Taiwan and that congressional resistance to the "abandonment" of the ROC would be intense.

As normalization grew closer, Taiwan's legislative supporters made no mystery of their readiness to intervene. Senator Barry Goldwater sought to introduce a resolution that would stipulate the necessity of the president receiving the advice and consent of the Senate before making any changes in the Mutual Defense Treaty. However, it was a bipartisan resolution (the so-called Dole–Stone amendment), stipulating that the president should consult with the Senate before making any changes in the treaty, which passed that body by a vote of 94–0 early in the summer of 1978, that clearly put the White House on notice (Harding 1992: 84).

The State Department pressed for greater consultation with Congress. However, as the management of normalization shifted to the White House, the process became increasingly secret, with only a small group privy to its details. Communications with China were over a White House backchannel, and access to communications was limited to a very few (US Department of State 2013a: 20). Concerns grew about the reaction to any leaks regarding the process, and consultation with Congress became less of a priority. In October, Brzezinski proposed to the president that advance notice of termination of the treaty should be, at least, given to the majority leader of the Senate, Robert Byrd. However, Byrd himself advised "against informing the Hill until the very last minute" – and this is precisely what the president did on the evening of the announcement on December 15, 1978, when he met with only a small group of Congress members while most of the body was on Christmas recess (ibid.: 556 and 1154).

This apparent snubbing of Congress, as well as the terms of normalization that gave notice regarding the Mutual Defense Treaty but failed to secure a pledge of peaceful resolution from Beijing, provoked a political outcry.[5] It was almost inevitable that, when the Carter administration submitted legislation – the Taiwan Omnibus Bill – intended to maintain "commercial, cultural and other relations with the people on Taiwan on an unofficial basis," Congress would revise it. In line with Christopher's discussions, this bill, hurriedly drafted by the White House, was concerned largely with maintaining the status of "the people on Taiwan" as a foreign state under American law and providing for the creation of a non-profit institution, the American Institute in Taiwan, to represent American interests.

Responding to the criticisms that the bill lacked security assurances, administration spokespeople argued that such language was unnecessary. The United States had made its interest in "peaceful resolution" clear during negotiations and would continue arms sales after 1979. Additionally, it was argued that the use of force would damage PRC economic ties with the West, that Beijing lacked the capacity to invade the island, and that Taiwan would maintain its defensive edge because of continued American arms sales. Finally, President Carter warned that he would veto any legislation that he deemed contrary to the terms of normalization.

Many in Congress were not impressed with either the argument or the threat. A wide range of proposals intended to deal with the consequences of normalization were put forth. However, the leading response was embodied in a joint resolution introduced by Edward Kennedy and Alan Cranston in the Senate and Lester Wolff in the House and drafted in close coordination with the administration. The sponsors sought a "legislative package" that would provide for "substantive continuity in the vital security sphere…on unofficial terms" in the same manner as administration legislation had provided for such continuity in "commercial, cultural and other relations." The goal would be "no more nor less than ... existing security commitments" (the 1954 treaty). The result was the Taiwan Relations Act (TRA), passed by Congress in April of 1979.

The TRA's statement of intent begins by asserting that "peace and stability in the [Western Pacific] area [including both the PRC and Taiwan] are in the political, security and economic interests of the United States and are matters of international concern," and that normalization had been based on the "expectation" of peaceful resolution of cross-strait differences. Any attempts to "determine the future [of Taiwan] by other than peaceful means" would be considered "a threat to the peace and security of the Western Pacific area and of grave concern to the United States." What is significant here is the contrast with the 1955 Mutual Defense Treaty, which identified hostility *toward Taiwan* as a danger to the "peace and security" of the United States. The drafters had intentionally defined the broader Western Pacific as in the American interest and an attack on Taiwan a lesser matter of "grave concern" within the context of those interests. With this broader area, the statement of intent called for the United States to maintain the "capacity to resist any resort to force or other forms of coercion that would jeopardize the security, or the social or economic system, of the people on Taiwan" and "to provide Taiwan with arms of a defensive character."

Under what conditions would that "capacity" be used to resist an attempt to coerce Taiwan? The TRA mandates that the president should inform Congress of the existence of "any threat to the security or the social or economic system of the people on Taiwan and any danger to the interests of the United States arising therefrom," and that they jointly "determine, in accordance with constitutional processes, appropriate action by the United States in response to any such danger." In other words, despite the language about grave concern, there is no *obligation* in the TRA to support Taiwan in the event of mainland hostility.

Regarding the sale of "defensive arms" to Taiwan, Congress also responded sceptically to the executive pledge to continue the sale of defensive weapons to Taiwan and inserted the commitment in the Act. However, by stipulating that such sales be made "according to procedures established by law," the Act weakened congressional influence by putting it in the passive position of approving or rejecting arms sales proposals put before it by the president as stipulated by the existing Arms Export Act.

These loose commitments of the TRA were the result of compromise among very different concerns in Congress that included the fear of entrapment by Taiwan in a cross-strait confrontation; the importance of preserving US–PRC normalization; and, most of all, a widely shared desire to affirm congressional prerogative. In the end, Congress created legislation that maintained enough of a commitment to Taiwan to satisfy concerns regarding the abandonment of an "old ally" but that also remained general enough to secure wide congressional support and avoid a veto. In short, the maintenance of ambiguity and flexibility in the American position on cross-strait relations that many would later criticize as a sign of weakness was an integral part of the TRA's political logic and the key to its passage with bipartisan support.

While the legislation was pending, China's foreign minister warned of the "great harm" that would be done to Sino-American relations by its passage and threatened a "necessary response." Deng Xiaoping commented that the passage of the bill placed a strain on the new relationship (US Department of State 2013a: 835–6 and 871, n. 4). The president sought to mollify the Chinese with his statement at the time of his signing of the TRA that it would be managed in "a manner consistent with our interest in the well-being of the people on Taiwan and with the understandings we reached on the normalization of relations with the People's Republic of China" (ibid.: 859).

The final year of the Carter administration was concerned with creating the infrastructure of the new relationship with Taiwan.

According to the first American representative in Taipei, Charles Cross, policy toward Taiwan in the post-normalization era would be guided by three principles: assuring that "actions concerning Taiwan do not impede relations with the PRC"; maintaining substantive relations with the island (especially economic); and helping to preserve stability by "maintaining the confidence of the people and leadership in the durability, reliability and profitability of the new relationship with the U.S." Indeed, Cross saw the "confidence" engendered by the final factor as an important element in contributing to the possibility of the peaceful settlement of cross-strait differences sought by the administration (US Department of State 2013a: 1125). It would soon be apparent that, from the perspective of Beijing, it would have the opposite effect. Hua Guofeng had warned Brzezinski to this effect, and Deng had angrily told Woodcock that, with American support, "Chiang Ching-kuo's tail would be 10,000 meters high." That Beijing was by no means done with the Taiwan issue in Sino-American relations would become apparent very soon.

RONALD REAGAN, TAIWAN, AND THE SHAKY LEGACY OF NORMALIZATION

During the 1980 presidential campaign, Ronald Reagan depicted normalization as the desertion of an old ally and a loss in the global battle against communism. He was opposed to the unofficial nature of American ties with the island and pledged to restore "officiality" to the relationship and use presidential "discretion" in applying the TRA. His intent was obvious. He would use the Act not to preserve the achievements of normalization, which had been its intent, but, rather, to dilute them. The Chinese, of course, were carefully following the campaign. In Beijing, Ambassador Woodcock reported to Washington in August of 1980 that "the Chinese have been provoked into reaffirming to us their right to liberate Taiwan by force of arms, a theme that has been muted since normalization" (US Department of State 2013a: 1133).

Thus, although he had backed off of his statements about reintroducing "officiality" into the relationship with Taiwan, when Reagan took the oath of office, both sides of the Taiwan Strait were anticipating a change in American policy – in Taipei, for the better, and, in Beijing, for the worse. The issue that immediately came to the forefront was the most fractious of them all – arms sales – in particular, consideration of the sale of a new fighter jet for Taiwan. The Carter administration had been pressed for the sale of a new-generation plane (an FX; the x simply meant the exact model was undetermined), rather than simply allowing the extended co-production of an older fighter already in use. In its closing days, the Carter administration decided to hold off a ruling on what kind of plane to make available and leave it to the next administration (US Department of State 2013a: 1159).

However, despite Reagan's campaign rhetoric and the presence of a strong pro-Taiwan group in his administration, the sale of a new aircraft was not certain. Alexander Haig, the incoming secretary of state, opposed such a sale. Haig, a former commander of NATO forces and an aide to Henry Kissinger, played on the anti-Soviet impulses of the new president and sought to create an anti-Soviet "strategic association" with China that would be sealed by the sale of "lethal, offensive" weapons to the Chinese. With Reagan's blessing, he set off for Beijing in June of 1981, only to learn the very same lessons that Kissinger and Brzezinski had learned (Tyler 1999: 296–300). Any talk of a global strategic alliance required an understanding on Taiwan – and, given the deep distrust of the Reagan administration, the going was rough. Foreign Minister Huang Hua told Haig bluntly: "U.S.–PRC cooperation cannot proceed without an understanding on TAS [Taiwan arms sales] which impinge on PRC sovereignty and are a 'grave threat' to development of U.S.–PRC relations" (Solomon 1999: 83–4).

Perhaps in an attempt to salvage the trip, Haig, without White House authorization, made public his approach to arms sales to China. This infuriated some in the administration and, while the secretary of state was in Beijing, Reagan reminded a news conference in Washington that he had not changed his "feelings" about Taiwan and would "live up" to the TRA, which "provides for defense equipment being sold to Taipei." To the infuriated Chinese, it seemed that the Reagan administration was seeking to use arms sales to Beijing as a means of gaining their acquiescence in parallel sales to Taiwan (Mann 1999: 121–2).

During 1981, the debate in the administration over the sale of a new fighter plane continued. Taking the initiative, the Chinese apparently made a decision that, rather than deal with the vagaries of the Reagan administration any longer, they would settle the arms sale issue once and for all. In September of 1981, Beijing put forth a nine-point plan for the unification of China which offered Taiwan "substantial autonomy under the sovereignty of the People's Republic" (Romberg 2006: 121–4; Tyler 1999: 110). Premier Zhao Ziyang then cited the nine points in a meeting with President Reagan the following month and repeated the mainland mantra that "Taiwan arms sales interfere in China's internal affairs and discourages Chiang Ching-kuo from negotiating" (Solomon 1999: 87). At the same time, Huang Hua proposed an agreement based on two points: that the United States would pledge that "arms sales to Taiwan will not exceed that of the Carter administration in both quantity and quality and [that] arms sales will be reduced year-by-year and completely stop." Haig apparently responded later that the president could "not accept a certain or fixed time frame for ending the TAS [Taiwan arms sales]"; but the process of negotiation had begun (ibid.).

By January 1982 increasingly ominous Chinese statements about the impact of arms sales on the relationship, and growing concerns in Washington over the impact of the FX issue on Sino-American relations against the background of Soviet pressure on Poland, led to a decision against supplying an advanced fighter to Taiwan, but rather extending the production of the F-5 (Romberg 2006: 127, n. 30). By the spring, Reagan had sent conciliatory letters to the Chinese leadership, pledging adherence to the concept of one China, welcoming the nine-point proposal and the statements of peaceful resolution, and stating that he would not allow "unofficial relations with the people of Taiwan to undermine" relations with China – almost the reverse of his earlier attitude. Beijing had once again made Washington's Taiwan policy the touchstone for good relations, and another president had concluded that there was no choice but to meet their demands.

Reagan's apparent willingness to accommodate Chinese demands became apparent with the announcement of the joint Sino-American Communiqué of August 17, 1982, and accompanying administration statements. The communiqué was ostensibly an agreement on arms sales – an issue that had not been "settled in the course of negotiations between the two countries on establishing diplomatic relations." However, in the manner of the Shanghai Communiqué, there were a number of joint and unilateral statements, among which was an American statement that the United States had "no intention of infringing on Chinese sovereignty and territorial integrity, or interfering in China's internal affairs," and a Chinese acknowledgment that it had "promulgated a fundamental policy of striving for peaceful reunification of the motherland." The key section on arms sales to Taiwan declared that

> the United States Government states that it does not seek to carry out a long-term policy of arms sales to Taiwan, that its arms sales to Taiwan will not exceed, either in qualitative or in quantitative terms, the level of those supplied in recent years since the establishment of diplomatic relations between the United States and China, and that it intends gradually to reduce its sale of arms to Taiwan, leading, over a period of time, to a final resolution. In so stating, the United States acknowledges

China's consistent position regarding the thorough settlement of this issue. (China–US 1982)

With this agreement, both sides announced optimistically that the final issue left over from normalization had been resolved. However, it was soon apparent that this was not the case. The Reagan administration, like previous administrations, was forced to rebalance its China policy to take account of the island's sensitivities as well as domestic political pressures.

Seeking to "ease the shock" of the announcement and to avoid any impression that Taiwan was being "sold out," the American representative in Taipei was instructed by the White House to deliver six assurances to Chiang Ching-kuo regarding negotiations with the Chinese. These were that the United States "has not set a date for ending arms sales to Taiwan; has not agreed to consult with the PRC on arms sales to Taiwan; will not play any mediation role between Taipei and Beijing; has not agreed to revise the Taiwan Relations Act; has not altered its position regarding sovereignty over Taiwan; and will not exert pressure on Taiwan to enter negotiations with the PRC." The day before the communiqué was agreed to, yet another message was sent to Chiang from the president, emphasizing that he "remains firmly and deeply committed to the moral and legal obligations in the Taiwan Relations Act"; that it was "United States policy to provide Taiwan with sufficient arms…to maintain a sufficient self-defense"; and that "the entire understanding on our [the US] part is predicated on the continuation of a peaceful policy on the part of the PRC." Chiang was assured that, if this peaceful policy changed, so would American assessments of Taiwan's self-defense needs. Further diluting the impact of the agreement with China, Taipei was authorized to make the six assurances public (Romberg 2006: 134; Feldman 2001).

The announcement of the communiqué caused a sharp reaction in Congress. Democrat John Glenn set the tone of the hearings when he charged that the administration had "discarded" the framework of the

TRA. In response, the Reagan administration "clarified" certain elements in the communiqué, asserting that arms sales could be increased in a crisis and that the agreement was premised on Beijing's commitment to peaceful reunification. Further elaborations of the American position were that the future level of arms sales would be adjusted for inflation; that the transfer of defense technology would not be included in calculating sales; that Taiwan had been given assurances (the "six assurances") regarding American policy; that weapons systems would be replaced by modern – not outdated – technology; and that the phrase "leading over time to a final resolution" specifically referenced the peaceful resolution of the mainland–Taiwan conflict, not some pledge that arms sales would end at a certain date. Between various glosses that were put on the agreement by the Reagan administration and the pledges given Taipei, it was obvious that there would be little change in American policy toward Taiwan (Tyler 1999: 327).

Of course, the Chinese interpretation was very different. It denied that it had committed in the communiqué to the use of only peaceful methods to solve the Taiwan issue and insisted that continued American arms sales were a violation of Washington's pledge to respect China's sovereignty. Further, Beijing interpreted the American statement that "it intends to reduce gradually the sales of arms to Taiwan, leading over a period of time to a final resolution," to mean that sales would be reduced and expire over time. Any effort by Washington explicitly to link arms sales to a peaceful unification policy by the mainland was rejected. As with the normalization accord, contestation rather than agreement had been the result of Sino-American negotiations regarding the Taiwan issue.[6]

CONCLUSION: AN UNCERTAIN LEGACY

Since the 1950s, unresolvable Sino-American differences over Taiwan had been both a major factor preventing the resolution of post-war tensions between the two nations and a cause of the crisis environment in the cross-strait area. In the decade from 1972 to 1982, both sides came to the negotiating table and sought to resolve the Taiwan issue in the interest of securing broader geostrategic goals. The outcomes of these negotiations in the form of joint agreements, congressional legislation, and unilateral pronouncements together form the foundation of the rules by which the United States and China have since attempted to manage their relations with Taiwan and with each other. However, as we shall see in subsequent chapters, this foundation has been a shaky one that has often complicated cross-strait tensions. The fundamental issues that created confrontation and deadlock in the Taiwan Strait remained unsettled.[7]

The most uncertain issue is the most basic one – the status of Taiwan. In 1972, although President Nixon had accepted Taiwan's status as part of China in his discussions with Zhou Enlai, all public statements by the United States since have evaded the issue. The Shanghai Communiqué merely stated: "the United States acknowledges that all Chinese on either side of the Taiwan Strait maintain there is but one China and that Taiwan is a part of China." In the last session with Deng Xiaoping, a contrite Woodcock tried to calm Deng by saying that President Carter would soon tell the American people "that there is one China and that Taiwan is a part of that one China." This was unlikely. When Woodcock reported on an earlier session with the foreign minister, he noted that the latter maintained that the United States had reaffirmed that "there is only one China in the world and that Taiwan province is a part of the People's Republic of China." In the margin, Carter had written: "We have not – stick to Shanghai language" (US Department of State 2013a: 611, n. 7).

The joint communiqué issued at the time of normalization continued the mutual obfuscation of the issue of Chinese sovereignty over Taiwan. It simply stated: "the United States of America acknowledges the Chinese position that there is but one China and Taiwan is a part of China." This was considerably less than the first of Nixon's five points. Indeed, it might have been less of a tilt toward the Chinese view than even the Shanghai Communiqué. It was certainly less than what Beijing might have expected given the mood that surrounded normalization.

This issue is further confused by the language used by each side. In the Shanghai Communiqué the Chinese translation for "acknowledge" had been "*renshi dao.*" In the Joint Communiqué of 1982 and 1978, the Chinese insisted on the term "*chengren*" – a term closer in meaning to "recognize." For some – including perhaps the Chinese who read the communiqués in their own language – it could be taken as recognition that Taiwan was a part of China. However, Washington has insisted that the English is controlling and that it does not constitute such recognition. Although the United States uses the term "one China policy" to refer to official relations with the mainland and not Taiwan, it is in no way comparable to the mainland's "one China principle" which asserts that Taiwan is a part of China.

Moreover, during these years the United States continued to act in a manner that, in Chinese eyes, continued the interference in China's internal affairs that had begun in the Civil War. Although the recognition of the PRC placed the ROC outside the category of a sovereign state, Washington still engaged in actions that treated it as such and had very little respect for Chinese claims of sovereignty over the island. It still reserved the right to treat Taiwan's "people" as an international actor for the purposes of domestic law and, even worse, to sell arms to them. The United States not only insisted that the two sides should resolve cross-strait differences peacefully, but, through the TRA, it legislated the possibility of American intervention in that process should the mainland's actions not be in accord with this preference.

Another reason for the weakness of the foundation constructed during these years was Sino-American differences over the identification of the operative texts and the nature of their authority. Thus, the Chinese reference three fundamental documents – the Shanghai

Communiqué, the Joint Statement at the time of normalization, and the August 17 Communiqué – as collectively forming the guidelines for handling the Taiwan issue. The United States cites these three documents but then adds a fourth – the Taiwan Relations Act. Moreover, in contrast to Beijing, which considers the TRA to be inconsistent with the other three, the United States takes the position that, because it is a law passed by Congress, the TRA is not only consistent but takes precedence over the communiqués.

It is apparent that the weak foundation of the rules for the management of the Taiwan issue in Sino-American relations results from the simple truth that the mutual compromise that took place during the process of normalization was circumscribed. The United States accepted China's three conditions. However, domestic political pressures and concerns regarding its international reputation dictated that Washington's concessions be accompanied by documents, from the TRA to Reagan's six assurances and qualifications subsequently attached to the 1982 communiqué, that together amount to the maintenance of much of the earlier relationship with Taiwan that was unacceptable to the mainland. Of course, neither had the mainland compromised its fundamental positions on issues such as the status of Taiwan, arms sales or the possible use of force to resolve cross-strait differences. In short, the frequent assertion that both sides "agreed to disagree" on basic issues relating to the Taiwan issue during the period after 1972 is simply wrong, not only in respect to arms sales but also in regard to most other key issues that separated them.

The irony is that Sino-American relations had been "normalized" in the same manner that the process of rapprochement had been begun in the Shanghai Communiqué – by means of parallel but often contradictory or differently interpreted statements on issues upon which consensus was impossible. This was not unusual given that the two nations, in 1972, were less than a year into unfamiliar waters after thirty years of intermittent conflict. However, normalization was a different case, where unresolved differences on very fundamental questions relating to each nation's relationship to Taiwan remained at the same time as the two were embarking on a far more complex new relationship.

There is a second similarity between the Nixon and Ford negotiations with China and normalization. The Nixon initiative had achieved some success despite deep differences because of a shared concern about the growing power of the Soviet Union. However, whatever common purpose this may have provided for the foundation of the Shanghai Communiqué was eroded by persisting unresolved differences over the Taiwan issue.

The normalization negotiations also took place against the background of a shared concern about the global behavior of the Soviet Union. Deng's rush to achieve an agreement and his repeated concerns that there should be no signs of discord at the time of normalization suggest he was looking over his shoulder at the Soviet reaction. However, as subsequent chapters will show, as was the case with the Nixon and Ford years, the dramatic impact of recognition would wither when the demise of the Soviet Union once more exposed Sino-American relations to the corrosive power of the still unresolved Taiwan issue.

To sum up, it is true that the normalization of Sino-American relations undoubtedly changed radically the dynamics of the cross-strait strategic environment. However, it neither eliminated the Taiwan issue as a divisive element nor permitted the disentanglement of the United States from the legacy of the Chinese Civil War embodied in cross-strait relations. Indeed, as we have seen, it simply recast the Sino-American relations on the issue of the Taiwan Strait triangle in a new form, with contested or uncertain rules that would provide a shaky foundation as cross-strait relations entered an entirely new and more complex stage as a result of the democratization of Taiwan.

4 | The Challenges of a Democratic Taiwan

In December of 1984, after they had agreed to the return to China of Hong Kong, Deng Xiaoping gave Margaret Thatcher a message for President Ronald Reagan. He suggested that the two sides work together to settle the Taiwan issue on the basis of the "one country, two systems" proposal he had offered Taipei in 1983 and which had just been applied to Hong Kong (Mann 1999: 153). The essence of this policy was the pledge that,

> after the reunification of the motherland, the Taiwan Special Administrative Region can have its own independence, practice a system different from that of the mainland, and its independent judiciary and right of final judgment need not reside in Beijing. Taiwan can retain its army so long as it does not constitute a threat to the mainland. The mainland will station neither troops nor administrative personnel in Taiwan. Taiwan's party, government and army departments are managed by Taiwan itself. The central government will reserve some seats for Taiwan … The systems can be different, but only the People's Republic of China can represent China in international affairs. (Wen 2009)

The request that Thatcher contact Reagan was consistent with Deng's long-held belief that the principal obstacle to reunification was the United States. If its "interference" could be removed, he believed, the common vision of one China shared by the KMT and the CCP could be the basis for reunification.

However, times were changing. In Taiwan, the era when the policies of an authoritarian government subject to the demands of an American ally set the tone of relations with the mainland with little concern for public preferences was ending. Democratization was already in motion in Taiwan, which saw the emergence by 1986 of an opposition party and the enfranchisement of islanders who, as we have seen in chapter 1, had very different historical experiences and future expectations. This was also the case with Deng's view of the role of the United States. With the KMT as a virtual client, dependent on its good will, Washington could exercise considerable influence. The enfranchisement of the people of Taiwan changed all that as well. Deng's proposal was, in short, out of touch with the reality of the 1980s and 1990s.

In this chapter, we trace the course of democratization, its impact on cross-strait relations, and the policy responses of the United States and China. With these factors in place, the scene for contemporary relations to be discussed in the last part of the book is finally set.

TAIWAN: A POLITICAL SYSTEM TRANSFORMED

Despite Washington's propaganda before normalization, the ROC hardly deserved the appellation "free China." It was a system where politicians who had come from the mainland dominated a single, unelected party that controlled the government. Any traces of democracy or rule of law that might have been present in the formal constitutional provisions were negated by provisions promulgated on the mainland during the civil war, which provided for an extensive military role in governance and restricted travel, assembly, speech, publications, etc.

By 1996, Taiwan's politics had been transformed. Single-party rule had ended. This was no longer a regime transplanted from the mainland that maintained the fantasy of the right to rule all of China. Only the people of Taiwan now elected the president and the entire national legislature. Moreover, the mainlander dominance of party and state offices had ended – a development dramatically symbolized by the fact that the president and chairman of the Kuomintang was now Lee Teng-hui, a Taiwanese veteran of the World War II Japanese army. At the lower ranks of the KMT, the shift in ethnic composition was dramatic. Between 1952 and 1993, the proportion of KMT party members of Taiwanese origin had grown from 26.1 to 69.2 percent; members of the ruling Central Standing Committee of the party from 0 to 57.1 percent; and members of the cabinet from 5 to 45 percent (Lin 1998: 271). Changes in Taiwan society were also evident. Civil society was alive with political movements. Moreover, writers and educators were free to explore the history and culture of Taiwan. Once almost a forbidden zone, Taiwan studies flourished, as did a growing sense of what came to be called "Taiwan identity" that valued the island's distinctive history and culture which KMT authoritarian rule had sought to denigrate.

Theorists have identified different factors to explain the democratic transition. Some focus on the role of elite politicians, others on social changes such as the growth of a middle class, while still others look at external factors such as the role of the United States. Although contextual factors are undoubtedly important, any explanation of democratization in Taiwan must begin with the actions of Chiang Ching-kuo, Chiang Kai-shek's son and his successor as authoritarian ruler.

Chiang had been at his father's side during some of the most repressive times of KMT rule on Taiwan. He was no democrat. He was a pragmatist, committed to the survival of the KMT. It is likely that this survival instinct motivated him to relax authoritarian rule, which, in turn, provided the legitimation and the foundation for the emergence of democracy after his death. The initial stages of his political liberalization began in the 1970s, when KMT legitimacy was challenged by China's growing international acceptance and American recognition.

At the same time, there was a growth in societal activism and the emergence of a loosely organized island-wide political movement that became known as the *dangwai* or "outside the party."

Chiang initially dealt with these challenges with policies that combined "selective repression and institutional liberalization." For example, elections that were previously only local were expanded to the whole island and designated as "supplementary" elections to the ROC legislative body, providing greater representation for the people of "Taiwan Province." However, at the same time, demonstrations by regime opponents were harshly dealt with and censorship remained in force (Chou and Nathan 1987: 278).

From 1985 until Chiang Ching-kuo's death in January 1988 the process of liberalization intensified.[1] In March of 1986, Chiang signaled that there would be a new round of political reforms. However, in September, the *dangwai* movement jumped the gun and established an opposition party, the Democratic Progressive Party (DPP). Rather than crack down on this violation of the existing law, Chiang ignored it and used the move by the opposition to press on with his survival strategy for the KMT. He announced he would end martial law which had been in effect since 1949 and legalize the formation of new political parties. Toward the end of his life, he spoke of the need to expand reform, transforming the KMT into an elective party and creating a new legislature elected only by the people of Taiwan. However, these goals were never realized before he died; they would be the work of his successor, Lee Teng-hui.

Lee was a native-born Taiwanese who had been educated in the United States as an agricultural economist. He had served as mayor of Taipei and governor of Taiwan before being tapped by Chiang to be vice-president in 1986. Although he was not part of the KMT mainland establishment, he was involved in much of the reform activity. When Chiang died, Lee succeeded to the presidency. However, he was considered to be an example of Chiang's tokenism, lacking the political base and the prestige of the party's old-line mainlander establishment, many of whom immediately set about planning his replacement.

In his first two years in office, Lee was cautious.[2] He secured his position within the KMT and stayed within the narrow, constitutional boundaries bequeathed to him by Chiang. Once this was achieved, he expanded the political arena to include the political opposition and elements of civil society to further strengthen his position. Lee courted the Taiwan native vote by abolishing some of the symbols of the mainlander rule of the past. For example, in 1991 the Temporary Articles which had suspended large parts of constitutional rule were revoked. The following year the advocacy of independence was decriminalized and the Garrison Command was abolished. Finally, on February 28, 1995, President Lee signaled an end to the decades-long cover-up of the 2.28 Incident when he formally apologized on behalf of the government.

However, the most concrete example of end of mainlander rule and the empowerment of the islanders was Lee's expansion of electoral politics. In 1991, the stolidly conservative and mainlander-dominated National Assembly was dissolved. Its successor, elected by the island's population, took its place and began the task of constitutional reform. The result was the first direct, popular election of the Legislative Yuan in 1992, followed by the island's governor in 1994. Finally, in 1996, Lee Teng-hui became the first elected president of the ROC. The Taiwan public had been given its voice in the island's politics and, more significantly, in cross-strait relations.

DEMOCRATIZING TAIWAN AND CROSS-STRAIT RELATIONS

Toward the end of his life, Chiang Ching-kuo came to believe that a new era of democracy was dawning on the mainland for which Taiwan could be a model, thus vindicating the KMT. To expose the mainland's people to a reformed Taiwan, Chiang liberalized cross-strait policy, allowing trade as long as it was indirect and did not involve contact with officials. Total trade suddenly grew, reaching more than US$2.5 billion before he died. At about the same time, the relaxation of rules regarding the export of currency facilitated investment in China, which in 1988 registered eighty projects valued at $100 million. This was followed by permitting travel to the mainland under the pretext of facilitating visits to their families by ageing KMT veterans. By 1988, the year Chiang died, almost a half a million Taiwan tourists had gone to China. Finally, the long-standing policy against coexistence with the mainland in international organizations was changed. In December 1985, the government announced that Taiwan "will not dodge or shy away from private organizations in which Communist China is a member" (Goldstein 1999).

These changes were an initial indication of how democratic reform on Taiwan was influencing mainland policy. The "hands-off" attitude toward cross-strait trade was a response to a business community that viewed the mainland as an important economic opportunity as a result of the unfolding post-Mao reforms. Moreover, since the majority of the small-business people flocking to the mainland were native Taiwanese, these initiatives were attempts both to gain their support and to blunt DPP influence and independence sentiment. Finally, the expanded foreign policy addressed the sense of global isolation that was undermining KMT support and giving credence to DPP independence sentiment. In the next decade, with Lee Teng-hui as president, the effect of democratization on cross-strait relations became more profound, affecting the fundamentals of Taiwan's approach to the mainland.

As we have seen, there was considerable consensus between Beijing and Taipei during the years of KMT autocratic rule. Both sides rejected any suggestion of independence for Taiwan. Both agreed that Taiwan had been returned to China after the war and that the current separation was a temporary result of a domestic civil war that would end with one side victorious. The fundamental difference, of course, was that the mainland saw itself as the successor Chinese government whose sovereignty extended to Taiwan, while the ROC presented itself as the still legitimate government whose sovereignty extended to the mainland.

Taiwan's democratization challenged this consensus in two respects. It did so, in the first place, because of the institutional implications of democratization. Amending the constitution to permit election of both legislative bodies and the president *only* by the people of Taiwan effectively eliminated the myth that the ROC represented all of China. Rather, it implied separation. After all, how could a government that laid claim to all of China elect its rulers from only one province? In addition, the elimination of the restrictions on the constitution necessitated by the communist "rebellion" required some redefinition of the existing adversarial relationship based on a domestic, civil war scenario. If the mainland was no longer a place ruled by the enemy or "rebellious bandits," what was it? And what was its relationship with the island?

Secondly, democracy would not only permit the expression of formerly forbidden topics such as independence and the creation of a separate "Taiwan identity" by drawing on the island's unique history both before and after World War II, it would also provide a previously disenfranchised population with the opportunity to select those who would be making policy regarding the status of Taiwan and its relationship to the mainland in light of these previously forbidden ideas. In a democratic Taiwan, preferences for mainland policy would go beyond a narrow circle of mainlander KMT leaders committed to a vision of a united Taiwan and China.

At the presidential level, the key to Lee's success in overcoming the resistance of the mainlander establishment within the KMT had been to seek support and to use political resources outside of the party organization. In this effort, the nativist DPP was a most valuable ally.

However, on cross-strait relations, Lee had to be more circumspect in dealing with the DPP. To embrace the DPP position – which, in 1991, had been a call for independence – would be too extreme for his supporters within the KMT, and yet, at the same time, there was political capital to be gained by exploiting the historically based, widespread distrust of the mainland and mainlanders that was now being freely expressed. Lee's solution to exploiting cross-strait issues for his political benefit in a democratizing society was to frame the National Unification Guidelines, which laid the principles for the future of cross-strait relations. These guidelines established a position that nominally embraced the reunification position of the KMT led by Lee but, by the nature of its timetable and demands on the mainland, also appealed to those who opposed it.

The National Unification Council established under the president's office in 1991 was responsible for drafting the National Unification Guidelines (Mainland Affairs Council 1991). These were based on the assumption that "both the mainland and Taiwan areas are parts of China" and that the goal of "forthright exchange, cooperation, and consultation" was "to establish a democratic, free and equitably prosperous China" in the future. It was proposed that "the timing and manner of China's unification" would go through three stages.

The "short-term" (immediate) phase would involve exchanges between the two sides, which would be monitored by organizations established to protect "people's rights and interests." It would also be a period when the mainland would be expected to implement economic reform and the rule of law while Taiwan accelerated its constitutional reform. Under the "principle of one China," all cross-strait disputes would be peacefully resolved and the two sides would "respect – not reject – each other in the international community." In the "medium term," "official communication channels" would be created on "an equal footing," and there would be mutual visits and consultations by officials from both sides. In addition, the two sides would "work together and assist each other in taking part in international organizations and activities." A final stage envisioned the establishment of a "consultative organization for unification" that would adhere "to the goals of democracy, economic freedom, social justice and nationalization of the armed forces, jointly discuss the grand task of unification and map out a constitutional system to establish a democratic, free, and equitably prosperous China."

This was a very dramatic reconceptualization of the nature of both cross-strait relations and the ROC that clearly played to the expanded political arena in Taiwan. The ultimate stage of unification was presented as the result of a negotiated process intended to protect the "rights and interests of the Taiwan people." Moreover, it was expected that the mainland would show respect for Taiwan in the international community, abjure threats of force to solve cross-strait issues, and transform its political system and economy along non-communist lines before unification would occur. These guidelines not only suggested that unification, rather than being a given, would be conditional, they also pushed unification far, far into the future.

At about the same time as the guidelines were drafted, two bureaucracies were established to manage relations with the mainland, now defined by Taiwan as separate "political entities." The Straits Exchange Foundation (SEF) was to be a private foundation to deal with the mainland on technical matters while maintaining an "unofficial status" (Goldstein 2007). It was often referred to as a "white glove" organization that allowed the government to manage cross-strait relations at a distance. However, it had government officials on its board, received government funding, and was supervised by a second new bureaucracy, the Mainland Affairs Council, a cabinet-level ministry closely supervised by the president.

Democratization also influenced the program of the DPP. With its calls for democratization basically co-opted by Lee's reforms, the party recast its mainland policy to make it *the* political issue separating the two parties and dividing society. The DPP had always presented itself as the party of the native Taiwanese, demanding democracy and "self-determination" for the islanders disenfranchised since the 1940s. This implied a call for independence. However, it was only in the 1991 elections to the National Assembly that independence became part of the party's platform. The result was a decline in DPP support, from an earlier 30 percent of the vote to 24 percent. When, a year later, it presented a more muted independence position during the first all-Taiwan election for the legislature, its support rose to 31 percent (Chang and Tien 1996: 36). With nearly one-third of the vote, the DPP had emerged from the democratization process as an important political force in Taiwan which sought to establish itself as a true Taiwanese party and the defender of popular interests vis-à-vis both the ruling KMT and the mainland authorities.

It wasn't long before the cumulative effect of these democratic changes impacted on Taiwan's cross-strait policy, prompting a radical change in the nature of relations.

MANAGING CROSS-STRAIT RELATIONS UNDER DEMOCRATIC CONDITIONS

Chiang Ching-kuo's policies toward indirect economic relations with the mainland could not have come at a better time for some in Taiwan's business community. Rising domestic costs, an appreciation in the value of the New Taiwan dollar, threats of protectionism in the West, and the growing environmental and labor movements on the island made coastal China, with its shared culture, low-wage labor, and lax legal regulations, an attractive investment site. Small, export-oriented, and labor-intensive manufacturers flocked to the mainland. By 1995, mainland trade represented approximately 10 percent of Taiwan's external trade.

Economics clearly led the way in the new era in cross-strait relations. This, by and large, was due to democratization. It was the result of the spontaneous and largely unregulated activities of the business community that created economic *faits accomplis*. Without question, the early trade ties with the mainland were governed more by the needs of business than by the calculations or preferences of government. Governmental attempts to regulate mainland trade and investment began in 1988 by designating permissible imports and exports. In 1991 procedures intended to limit investment were promulgated. However, in a democratized Taiwan, it was difficult to control the "mainland fever" that was gripping the island's business community. The authoritarian government of old was gone. Democratization had both weakened the regulative capacity of the state and made it susceptible to outside pressures (Leng 1996).

With the economic relationship growing, there was a need to deal with certain "functional" issues relating to the growing exchanges across the strait. And so, in April 1993, "unofficial" representatives of both the mainland and Taiwan met in Singapore. These were the highest-level talks between the two sides since the end of the civil war and the first significant encounter between the mainland and a democratizing Taiwan.

In preparatory negotiations, which took place in Hong Kong in 1992, the first issue to come up was the nature of the relationship between the two sides – specifically, whether these talks were of a domestic or an international nature. The mainland was insistent that the talks should take place on the basis of the "one China principle," without any suggestion of "two Chinas" or "one China and one Taiwan," and certainly no mention of two "political entities." Taiwan presented much vaguer formulations based on the revised positions of the Lee government.

A negotiated agreement could not be reached after several meetings. The solution was an exchange of parallel statements in which each side stated its readiness to express orally its position on "one China," although it was clear that each side had a very different conception of the "one China" it was talking about. The Taiwan side indicated that it would adhere to the National Unification Guidelines as well as an August 1991 statement of the National Unification Council, which asserted Taiwan's sovereignty and equality and took "'one China' to mean the Republic of China (ROC), founded in 1911 and with de jure sovereignty over all of China. The ROC, however, currently has jurisdiction only over Taiwan, Penghu, Kinmen, and Matsu. Taiwan is part of China, and the Chinese mainland is part of China as well." The PRC's statement avoided defining "one China." It simply asserted that "Both sides of the Taiwan Strait adhere to the one China principle and are making efforts toward national unification. However, the political meaning of 'one China' is not involved in the Cross-Strait functional talks."

This exchange of statements would later become known as the "1992 Consensus" of "one China, different interpretations" (Wachman 2007: 7; Kan 2011: 46). However, it was a very thin consensus indeed. The Taiwan side's "one China" preserved the sovereignty and equality of the still existing ROC, while the mainland obviously saw itself as the successor Chinese government. However, Beijing apparently conceded that, as long as there was a veneer of "one China" over the merely "functional" talks, the contradictory nature of the statements would be acceptable.

The impact of Taiwan's democracy on the initial encounter with the mainland went beyond competing differences regarding the status of the two sides. The political divisions that had emerged during democratization shaped the preparations for, and management of, the talks. Differences between the "mainstream" and "non-mainstream" factions within the KMT divided the Taiwan delegation. At the same time, the DPP used every legislative tool at its disposal to intervene in planning for the meetings and sent a delegation to Singapore to protest at them.

Although the talks ultimately went forward, the DPP succeeded in gaining assurances from the government negotiators that narrowed its agenda (Goldstein 2007).

The growing influence of the DPP as the voice for Taiwanese identity and sovereignty was also reflected in its gains in the 1992 election to the Legislative Yuan. Lee Teng-hui began to co-opt these issues to benefit his candidacy. In 1994, there were two indications of this. The first was an interview with the Japanese writer Ryotaro Shiba, in which Lee spoke of the "grief" of being born on the island and depicted the KMT as a "foreign power" that had come "to rule" the Taiwanese. He suggested that unification with the mainland would bring another kind of mainland oppression that would be met by the people of Taiwan with another 2.28. Presenting himself as a Moses figure who would lead his people out of oppression, Lee spoke of the need for education in the history of Taiwan and of his aspiration to "build a nation state and a society for 'the public'" (Wachman 2007: 11).

As for the current status of Taiwan, Lee was cautious but also pro-vocative. He characterized Beijing's view that the island was a province of China as a "weird fantasy," since the mainland and Taiwan have "different governments." He stopped short of elaboration on this theme, saying, "I can only say this for now." A month later, an ROC White Paper drafted under his supervision elaborated on Lee's interview in even more provocative language (CSIS 1994). After asserting that "one China" referred only to China as "a historical, geographical, cultural, and racial entity," the statement maintained: "It is an incontrovertible historical fact that the ROC has always been an independent sovereign state in the international community since its founding in 1912" and declared that "the two sides should coexist as two legal entities in the international arena." Finally, the White Paper reminded Beijing that Taiwan was now a democratic society containing both "integrationist and separatist ideas" and that, while the government favored the former view, it would be guided by popular opinion that would react negatively to the use of "high pressure tactics."

The White Paper was related to another plank in Lee's developing presidential platform – achieving the international standing and respect that many Taiwan citizens felt they deserved. He began pursuing what came to be known as "flexible" or "pragmatic" diplomacy. Taipei aggressively sought to expand the scope and visibility of its foreign policy to include not only entry into the United Nations (a DPP demand co-opted by Lee) but also Lee Teng-hui's "vacation diplomacy," during which he traveled to Indonesia, the Philippines, and Thailand on "private visits" – ostensibly to play golf.

However, as the first popular presidential election approached, Lee was looking beyond Southeast Asia to practice his "vacation diplomacy." In the spring of 1994, he was to attend presidential inauguration ceremonies in Costa Rica. He requested a refueling stop in Hawaii and, perhaps, a round of golf. The US State Department at first was inclined to refuse any stop. However, it later approved a brief stop during which Lee would be escorted to a small VIP lounge. Lee considered this unacceptable and refused to leave the plane.[3]

His failure in Hawaii led to an even greater opportunity for Lee to bolster his political standing for the 1996 elections by enlarging Taiwan's international visibility. Cornell University, where he had earned his PhD, had invited him to speak at an alumni reunion in June 1995. The legacy of the Hawaii trip and the Republican control of both houses of Congress created a sympathetic audience for the request. Although the secretary of state had assured the Chinese that a visa was unlikely, a sense of the Congress resolution supporting a visit by Lee passed by 97–1 in the Senate and 360–0 in the House. President Clinton approved a visa. However, efforts by the State Department to keep the visit low-key were fruitless. Lee was determined to use the trip to highlight the recent changes in Taiwan's domestic and international situation ("the Taiwan experience") and, of course, his role in making them possible. The speech at Cornell lauded Taiwan's democratic freedoms, suggested that the example of Taiwan might "help" reform in China, and expressed the hope "that all nations can treat us fairly and reasonably, and not overlook the significance, value and functions we represent" (Lee 1995).

TAIWAN'S DEMOCRATIZATION: THE MAINLAND OBSERVES

The late 1980s to the early 1990s was a relatively tranquil period in cross-strait relations. As we have seen, economic relations were expanding and the 1993 meeting in Singapore was a truly historic moment. However, as democratization unfolded in Taiwan, mainland commentators became increasingly concerned about its implications. Almost from the very beginning, analysts suggested that the greater the level of democratization, the more unsatisfactory Taiwan's cross-strait policy would become. There were, of course, the expected statements about how reform was liberating the Taiwan people from KMT oppression and reducing hostility toward the mainland. Moreover, a government subject to popular pressure was assumed to be more vulnerable to demands from economic interests to promote relations with the mainland. However, almost from the beginning of the democratization process, commentaries in the mainland press reflected uneasiness that democratization was promoting independence.

The most obvious reason for this linkage was the emergence of the DPP. The party's call for Taiwan's independence, its ability to organize and draw on the islanders' historical grievances, and the return of some of its most radical leaders from abroad were the most obvious examples of democratization's deleterious impact on cross-strait relations. However, the DPP was not alone in using political reform to frustrate mainland reunification efforts. Throughout the early 1990s, mainland observers contended that the sudden growth of the DPP and independence sentiment in general were not simply the spontaneous consequences of the newly won freedom of speech and organization. They were due to the support of some in the ruling KMT who were favoring a political agenda intended to frustrate reunification (Roundup reviews 1989 political situation 1990: 38; Liaowang views "independent Taiwan" 1990: 67).

Beijing would accept none of this. It contended that the ROC had ceased to exist as a political entity in 1949. The PRC was the successor government of China, and Taiwan was a subordinate part of its sovereign territory, as stipulated in the wartime agreements. To accept the conception of an equal "political entity" as the basis for negotiations was impossible. The same assumption about the status of Taiwan shaped Beijing's view of pragmatic diplomacy and its reaction to peaceful reunification. Attempts by Taiwan to seek admission to the United Nations or to gain greater global recognition were seen as reinforcing constitutional separation from the mainland by creating "two Chinas" or "one China and one Taiwan." On peaceful reunification, the mainland's response was to reiterate that peaceful settlement was its basic policy but that, as the only legal government of China, it reserved the right to use force against those who sought independence or foreign countries that interfered with the unification process.

Until 1994, Beijing focused on the DPP as the political force actively seeking independence, while the KMT's constitutional tinkering was seen as currying favor with independence forces and placing conditions on cross-strait negotiations in order to maintain the status quo. In that year, Beijing's view of the KMT, and especially of Lee Teng-hui, hardened, and Lee became the focus of mainland frustration over the lack of progress toward reunification. This process intensified during 1995 as he was preparing to take his conception of Taiwan and its relation to the mainland to the polls (Renmin Ribao views Taiwan "one China" stance 1994: 94).

In January 1995, Jiang Zemin, general secretary of the Communist Party, made his first important foray in cross-strait politics by proposing an eight-point program that offered negotiations. However, it rejected the essentials of the new Taiwan position by including the necessity of acceptance of the one China principle, a refusal to abandon the option of force to oppose independence, and opposition to pragmatic diplomacy. In April, Lee dismissed Jiang's offer with his own six-point program, which reaffirmed his concept of two political entities and called for the renunciation of force while suggesting that some international organization might be a venue for talks.

With the speech at Cornell, Lee definitively established himself as the lightning rod for mainland frustration in its dealings with a democratized Taiwan. Commentaries on the speech depicted it as the final stage in Lee's evolution to the out-and-out advocate of independence. It was charged that he was now working in alliance with the DPP to achieve independence (Li's "lining" means creating "two Chinas" 1995: 56; Li's "political scheme" 1995: 82–7).

Mainland commentaries also identified the United States as Lee's partner. One argued that Lee was "acting at the United States' direction" because Washington was attempting to sabotage cross-strait relations (Commentary warns against Li Teng-hui's visit 1995: 50). With resistance to mainland blandishments growing in democratic Taiwan, Beijing instinctively reacted as it had since the 1940s by blaming American support for the ROC's behavior. By the spring of 1996, it was apparent that the newly democratized Taiwan was becoming the tail wagging two dogs. The beginning of direct cross-strait relations was intensifying, rather than reducing, the significance of Sino-American relations in the area.

THE UNITED STATES AND CHINA: THE TAIWAN ISSUE REDUX

After the 1982 arms sale communiqué, and for the remainder of the Reagan administration, Sino-American relations entered a stable period. Cross-strait relations seemed to be moving in a positive direction as economic relations flourished. The June 1989 events at Tiananmen ended this honeymoon period in Sino-American relations even as it enhanced Taiwan's image in the United States. As the mainland seemed to be regressing politically, democratic reforms were actually turning the ROC into what Washington had claimed it was in the 1950s – a kindred democracy in Asia. Although President George H. W. Bush sought to maintain some momentum in Sino-American relations despite the events at Tiananmen, his decision to sell advanced fighter planes (the F-16 A/B) to Taiwan in September of 1992 was consistent with the changed image.

As we have seen, since the last years of the Carter administration there had been an awareness of the need to upgrade Taiwan's air force. By 1992, Taiwan's need was even greater, but still greater was George H. W. Bush's need to carry Texas, where the planes were manufactured, in the 1992 presidential election. The size of the sale clearly violated the 1982 arms communiqué. However, domestic political imperatives trumped past commitments to China. Bush won Texas but lost the election. His China policy had been harshly criticized during the campaign by Bill Clinton. When the latter took office, he was under pressure from Congress to link Most Favored Nation status with an improvement of the human rights situation in the PRC. At the same time, there was growing pressure from Taiwan's congressional supporters to enhance the relationship with Taiwan.

Congressional activism increased in 1994, when a conservative Republican majority took control of both houses of Congress, dramatically changing the political context for United States policy in the strait. The first indication of this was the granting of a visa to Lee Teng-hui in the spring of 1995 discussed above.[4] At this time, cross-strait relations were entering a volatile period, as democratization on Taiwan had unleashed political trends directly challenging mainland interests, and, despite normalization, the United States continued to aid Taiwan. Lee's Cornell speech symbolized the confluence of unfavorable trends in the strait policies of both Taipei and Washington. China's leadership concluded that something had to be done.

In July 1995, after canceling a cross-strait meeting because of Lee's Cornell visit, Beijing announced that there would be missile tests in the open seas less than 100 miles from northern Taiwan. Despite this, Lee Teng-hui persisted in his defiant rhetoric, while the response of the United States was very low-key. However, the American attitude changed as signs indicated that Beijing was planning to resort once more to military demonstrations, in an attempt to affect legislative elections scheduled for December 1995 and the presidential election of March of 1996.

The legislative election was important because it would be the first major test of the pro-mainland New Party, which had split off from the Kuomintang during the summer of 1993. A week before the election, the mainland held a large-scale military exercise involving a simulated amphibious invasion. However, its primary purpose had been to send a message to the Taiwan voters – and apparently it had some success. The New Party received 13 percent of the total vote and twenty-one seats in the legislature. With this apparent success, planning in Beijing proceeded for another round of threatening exercises to influence the presidential election the following March. The mainland announced that it would conduct three rounds of military exercises that would end the day before the Taiwan elections. Beginning with the launch of missiles in the seas to the north and south of Taiwan, the exercises would then move to live-fire drills in the Dongshan area of Fujian, opposite Taiwan, and conclude with an amphibious invasion of an island off the coast of China. The intent to intimidate was obvious.

In contrast to the earlier exercises, these were taken very seriously by the Clinton administration, concerned that an accidental conflict might draw in the United States. However, the exercises also attracted the attention of the Republican Congress, which took the occasion of the growing tensions in the Taiwan Strait to assert its prerogative under the Taiwan Relations Act to shape not only the executive's cross-strait policy but also its relations with China. Invoking the TRA, voices were raised in Congress demanding a robust response to China's actions. From this point on, policy toward the strait area would become entangled in American domestic politics in a manner reminiscent of the 1950s and to an extent not seen since the Carter administration.

However, the Clinton administration preempted political pressures from Congress by taking the initiative. After a visiting Chinese official was admonished by high administration officials, using language drawn directly from the TRA, and missile firings continued, two aircraft carrier battle groups (the largest American military deployment in the area since the Vietnam War) were ordered to the area. The next day, the Chinese announced the suspension of missile tests. Although a tense atmosphere would persist through Taiwan's presidential elections, the confrontation began to wind down.

In the end, the presidential election in Taiwan went off without incident. Lee campaigned vigorously as a politician who would not be intimidated by the mainland's threats and personal attacks. In the end, he received 54 percent of the vote (enough to avoid a runoff), with the DPP candidate receiving 21 percent. In other words, as observers noted immediately, this was not only a personal victory for Lee, but 75 percent of the total vote had gone to two candidates opposed to moving toward unification. It was a sharp rebuff by a democratic Taiwan to mainland policy.

It is also generally recognized that the danger of actual conflict was less than might have appeared. The United States and China stayed in close communication, and both exercised considerable restraint. However, despite its somewhat controlled – some would say, even choreographed – nature, the events of 1995–6 had a profound effect on the policies of all three actors in the triangle. The initiation of the three rounds of military exercises was clearly intended to signal Beijing's dissatisfaction with the direction of Taiwan politics and its loss of confidence in Lee Teng-hui as a negotiating partner. How much the Chinese leadership expected to shape the electoral outcome is uncertain. Still, it is evident that, if they were seeking to use threats to influence the election, they had plainly failed – or even prompted an opposite result.

A second audience for China's use of force was, of course, the United States. As noted, a fundamental belief of Beijing's calculus was that American support stiffened Taiwan's resistance to mainland blandishments. The displays of Chinese force were intended to affect American calculations as much as Taiwan's public. In this respect, as we shall see, there appears to have been greater success.

Analysts in the West differ over the impact that the crisis might have had on the balance of power within the Chinese leadership. Some argue that, throughout the 1990s, the military had been pressing Jiang Zemin to pursue a firmer policy toward Taiwan and that the exercises were an indication of their influence on this relatively new Chinese leader. Whether or not this is the case, the outcome of the show of force certainly did benefit the military. By triggering the deployment of two aircraft carrier groups, it had demonstrated the power of the American military in the area, while highlighting the need for a significant increase in military capability to respond to future provocations from Taiwan and interventions from the United States. As we shall see in chapter 8, beginning in the mid-1990s, the Chinese military undertook a massive build-up in the strait area that tipped the military balance, posing problems not only for Taiwan but also for the United States.

Finally, the confrontation had important effects on American policy toward both Taiwan and China. On the former issue, statements made to deter China were without precedent in the post-normalization period. Previously, policy had been that of "strategic ambiguity," leaving open whether the United States would intervene in a strait crisis. This uncertainty was, as we have seen, also written into the TRA. However, the dispatch of the aircraft carrier groups suggested an end to ambiguity and an expression of a willingness to defend Taiwan. Moreover, the congressional testimony of administration officials explicitly declared a readiness to come to Taiwan's defense should it become necessary (Goldstein and Schriver 2001: 156–8).

The confrontation also alerted some in the administration and Congress both to Chinese sensitivities on the Taiwan issue and the problematic state of Sino-American relations. The Clinton administration had gotten off to a bad start with China when it linked Most Favored Nation treatment to human rights and had done little to repair relations since. By the late spring of 1996, initiatives from the executive branch were underway to ameliorate that situation. The reaction of Congress was mixed. Some members seemed to have gained a greater appreciation of China's sensitivities on the Taiwan issue. Many expressed surprise that Lee's visa had triggered such a strong PRC reaction and indicated that they would not have supported it had they better understood the likely consequences. However, the mainland's attempts to intimidate a "democratic Taiwan" fed the growing partisan divide separating the administration from its Republican opponents. For the remainder of Clinton's presidency, his opponents in Congress would make good use of legislative power in their attempts to frustrate any improvements in Sino-American relations.

It is difficult to say how the events of 1995–6 affected Taiwan politics. Despite the bravado of Lee Teng-hui, the demonstrations of force did affect the population of Taiwan, as demonstrated by the precipitous fall of the stock market at each crisis point in the spring. However, the size of Lee's victory was considered to be a clear rebuke to the Chinese. Moreover, in his inaugural address two months after the election, Lee was defiant. He spoke several times of "popular sovereignty" (i.e., sovereignty based on the vote of the people and not any historic claims) as the basis for the right of the ROC to be considered a "sovereign state." He repeated the concept of "two jurisdictions" and pledged to continue to seek "respect" and "room for existence" in the international arena. Finally, in response to the recent events, he pledged that Taiwan would "never negotiate under threat of attack" (Lee 1996).

However, probably the most significant fact was that the events surrounding the election in March had put the issue of cross-strait relations to a popular vote. For the first time, the population of Taiwan had the opportunity to express its preference, and the result was overwhelming support for the two parties associated with continued separation from the mainland. The people of Taiwan had signaled their participation in cross-strait relations. It would increase in the decades ahead.

UNEASY CALM IN THE STRAIT: 1996–2000

For two years after Lee's Cornell visit, cross-strait relations were stalemated. The mainland consistently dismissed his statements of commitment to unification as rhetorical cover for his quest for independence. Moreover, Beijing kept up its drumbeat of opposition to Taiwan's application for admission to the United Nations and to its attempts to expand the island's international profile. Most important, there was little movement away from its insistence that Taiwan's place in "one China" be based on the principle of "one country, two systems," which assumed Taiwan's subordination to the mainland. In addition, the mainland showed very little readiness to be flexible on the "one China" question in discussing preconditions for talks. The result was that the "unofficial" cross-strait talks atrophied, and Beijing sought to broaden its influence on Taiwan beyond this channel by cultivating people-to-people contacts with business figures, local officials, and more unification-oriented politicians.

In Taiwan, the election of 1996 had demonstrated the domestic political value of defying mainland demands. Lee's government continued to assert that it was a "sovereign" and "independent" political entity and to insist on a "one China" position, based simply on history and culture, that in no way diminished its claim of being the equal, not the subordinate, of the PRC. The administration also persisted in its efforts to increase Taiwan's international profile. At the same time, it sought to strengthen Taiwanese identity. Lee Teng-hui's frequent reference to the island being a "community of shared fate" was intended to move in the opposite direction of earlier KMT efforts, toward building Taiwanese identity that could unite both *bensheng* and *waisheng*. Research into the history, literature, and culture of Taiwan flourished. Under the rubric of "Know Taiwan," the school curriculum was revised with these subjects centrally placed. The impact on popular attitudes was striking. A survey showed that, between 1992 and 1999, the number of respondents identifying themselves as Taiwanese had gone from 17.6 to 36.9 percent (Election Study Center 2015a).

At the same time, American policy toward both Taiwan and the mainland was undergoing significant changes. The 1996 confrontation had been a turning point in the Clinton administration's policy.[5] Congress had been placated by its strong response as well as by the apparent suspension of strategic ambiguity in defense of Taiwan, but these successes ironically only increased pressure on the administration. The security aspects of the Taiwan Relations Act had been given unprecedented public exposure, with the role of Congress in cross-strait policy highlighted. The president's legislative opponents proceeded to use the TRA as an instrument to influence future China policy. The occasion for congressional intervention was the renewal of Sino-American summitry as the president met with Jiang Zemin in an orchestrated pair of meetings in October 1997 in Washington and in June 1998 in Beijing.

Before the first summit, resolutions were introduced in Congress calling on the president "to make clear" to Jiang the American commitment to the security of Taiwan mandated by the TRA and to repeat the expectation, present at normalization, that cross-strait differences would be resolved peacefully. More pointedly, the Republican-controlled House passed legislation calling for the provision of antiballistic missile (ABM) systems to Taiwan and the island's possible inclusion in a theater-wide ABM system – an obvious complication for the developing rapprochement with China.

Before the June 1998 summit, rumors were circulated regarding the possibility of a "fourth communiqué" that would adversely affect the interests of Taiwan. Congress held hearings and administration officials pledged that there would be no change in policy related to Taiwan. However, in Shanghai, President Clinton articulated what would become known as the "three noes": no support for Taiwan independence, no support for "two Chinas" or "one China and one Taiwan," and no support for Taiwan's membership in any international body that requires statehood. These policies had been articulated previously, but the fact that the president expressed them in Shanghai only exacerbated his conflicts with congressional opponents who saw further evidence of appeasement at the expense of Taiwan.

Congressional intervention in China policy – and relatedly in cross-strait policy – was reaching a level unprecedented since normalization. And coincidentally, as the twentieth anniversary of the TRA approached, Congress acquired a powerful new bureaucratic ally – the Department of Defense.

The TRA had mandated that the military maintain the capacity to resist any resort to coercion in the area. Some in the Pentagon argued that the recent Chinese build-up required that greater resources be allocated in order to comply with that mandate. Moreover, the lessons of 1996 had demonstrated poor coordination between the two militaries. As one senior defense official remarked, "We were almost standing shoulder to shoulder with Taiwan in the conflict, and we knew far less about them than we knew about the PRC." A convergence of interests thus developed between Taiwan's supporters in Congress and some in the Defense Department. With information provided by the Pentagon, congressional supporters of Taiwan used the TRA to intervene directly in the arms sales process, demanding that specific packages be provided to the island.

As the Clinton administration and Congress struggled over the balance of influence in setting policy, Lee Teng-hui sparked another crisis in cross-strait relations. Speaking on the German radio station Deutsche Welle, Lee again denied mainland sovereignty over the island, asserting that it belonged to the people of Taiwan who elected their government. He added that, for this reason, cross-strait relations were a "state-to-state relationship or at least a special state-to-state relationship" (Lee 1999).

Beijing's reaction was immediate. A visit by the head of the Association for Relations Across the Taiwan Straits (ARATS) to Taiwan was canceled and the talks with Taiwan were suspended. Beijing insisted that it would not restart the talks until Taipei was ready to "return to the position of the 'one China' principle." The personal attack on Lee that followed was ferocious even by post-1996 standards. Washington's reaction was to deny immediately any association with the statement and to send representatives to Beijing. Relations with China were already strained over issues ranging from campaign finance to spying, and there was some concern that this might be a cause for a repeat of 1996. President Clinton called President Jiang Zemin to assure him that the United States held to its "one China policy" and was not supporting Lee's position.

Of course, the timing of the mainland threats could not have been better for the supporters of Taiwan in Congress and the Pentagon, who expressed their support. In the fall, a Taiwan Security Enhancement Act (TSEA) was introduced which included a paragraph in the findings section declaring that it would be "in the national interest for the United States to eliminate ambiguity and convey with clarity continued United States support for Taiwan, its people and their ability to maintain their democracy free from coercion and their society free from the use of force against them."

The revised TSEA passed the House by an overwhelming 340 to 70 in February 2000. The timing was not coincidental. It was the eve of the presidential elections in Taiwan. Beijing had expressed serious concern over its outcome, highlighted by a finger-shaking warning from the prime minister to the people of Taiwan (see chapter 5). The TSEA debate on the floor of the House suggests that many of its supporters sought to deter the mainland from a repeat of the 1996 election confrontation. However, the election passed peacefully, and the TSEA was never put up for a vote in the Senate. The new Taiwan government felt that it was not a favorable time for its passage, and, as Lee left office to be succeeded by the victorious DPP candidate, Chen Shui-bian, many in the Congress feared that it might provoke rather than deter the mainland.

CONCLUSION: THE MORE THINGS CHANGE . . .

At the beginning of the twenty-first century, there was much that had changed. In the first place, the democratization of Taiwan had created a pattern of cross-strait relations quite different from those that had prevailed during the previous fifty years. In contrast to the earlier period, the contacts between Taiwan and the mainland had become the primary driver of the strategic environment in the area. However, the mainland's relations with the United States still remained closely linked to its policies across the strait.

For the leadership in Beijing, reunification must have seemed a distant goal by the end of the 1990s. Despite overwhelming global recognition of its status as the successor government in China and the severing of official ties between Taiwan and the United States, the PRC was as far from realizing this objective as it had been fifty years earlier. In fact, one could argue that it was now even further away, since the consensus of "one China" that was shared with the KMT had evaporated with democratization. Locked into a policy of "one country, two systems" and a view of the island that placed it in a subordinate position in "one China," the mainland had little to offer the people of Taiwan who had recently found their political voice. By the end of the twentieth century, with cross-strait tensions growing on account of perceived independence sentiment on Taiwan, and following the dispatch in 1996 of two aircraft carrier groups to the Taiwan area, which forced the PLA to assume that the United States would participate in any cross-strait military confrontation, Beijing began to build a military capacity intended to deter movement toward that goal, or, if need be, to coerce Taiwan to accept unification even if the US were to become involved.

Frustration with Taiwan's resistance to mainland blandishments was, of course, linked to American support. From the Chinese perspective, since the end of World War II, American policy had been to deny Taiwan to China. Deng's prediction that the American connection would encourage the island's leaders to resist the mainland appeared to have been confirmed by Lee Teng-hui's behavior. Mainland declarations regarding the centrality of Taiwan in Sino-American relations and its designation as a "core interest" of China were intended to indicate the belief in Washington's centrality in relations with China.

In the United States, as was the case in Taiwan, domestic politics had become a driver in policy toward China. Bill Clinton had run for the presidency on a platform sharply critical of Beijing, and in the early years of his administration he followed through on this campaign rhetoric. When the administration sought to repair the relationship in the mid-1990s, a hostile Congress in an environment of growing concern regarding the "China threat" took up the cause of Taiwan, providing further provocation to China and reinforcing Beijing's doubts about American motives.

And so the United States found itself in the same position in which it had been more than fifty years earlier – caught in the middle of the two enemies in the Chinese Civil War. Since 1950, Washington had sought to disentangle itself from that position, only to become more deeply enmeshed as it moved from a position of support for one side to attempting to maintain relations with both. The impact of this policy was, as we have seen, to weave a web of ambiguous or mutually contradictory pledges to both sides that served not only to further entangle the United States in that conflict but also to weaken its credibility on both sides. Moreover, Washington's efforts to manage in this difficult position were complicated by Taiwan's democratization. Although at times the actions of the Kuomintang threatened to entrap Washington in a conflict with the mainland, for the most part it could be controlled, and American policies toward China could be pursued with little consideration of the KMT's preferences. This would not be the case when dealing with a government on Taiwan that had to balance domestic political necessity against American demands.

Cross-strait relations had entered a new and more complicated era, as would be evident very soon after the new DPP administration took power in Taiwan in 2000. However, in responding to this changed and potentially dangerous era, the United States and China were proceeding from a policy foundation of old assumptions and unsatisfactory agreements.

5 | Period of High Danger ───────

In March 2000, Chen Shui-bian became the second popularly elected president in the history of Taiwan. Chen, a former mayor of Taipei, had been the clear beneficiary of a split within the opposition that resulted in the candidacies of two Kuomintang heavyweights, Lien Chan (the former vice-president and the designated candidate) and James Soong (the former governor of Taiwan running as an independent). Although he garnered only 39 percent of the vote, Chen's victory was the stuff of great political drama.

The victorious DPP was less than twenty years old. It was a voice of the native Taiwan population's grievances against the mainlander government of the KMT and demands for democratization. The party had been at the forefront of the effort to overcome mainlander domination through democratization and, as we have seen, during the late 1980s, had begun a drift in favor of independence that culminated in its 1991 platform for the National Assembly elections. However, following its poor showing in that election the party made a tactical retreat, and in 1999 the national party congress passed a resolution asserting that Taiwan was already "a sovereign and independent country...named the Republic of China under its current constitution." Moreover, the resolution held that "[a]ny change in that status quo" required a plebiscite "by all residents of Taiwan" (DPP 1999).

The document was an obvious attempt to balance the need to retain the loyalty of the party faithful with that of appealing to a wider slice of the electorate. In addition, Chen's statements during the campaign were relatively moderate, and there was some optimism that a cross-strait crisis might be averted during his tenure or even that the strained relationship would be eased, since his party might have the credibility to make a deal with the mainland.

However, Beijing's attitude during the campaign offered very little promise of this. The past history of Chen and the DPP had not been forgotten and, apparently ignoring the lesson of the presidential election of 1996, as the elections neared, Beijing once again adopted a threatening posture. In February 2000 a White Paper entitled "The One-China Principle and the Taiwan Issue" was released by the PRC (Taiwan Affairs Office 2000). While restating the basics of China's position, it made one notable addition to the conditions under which force might be used against Taiwan: "if the Taiwan authorities refuse, sine die [indefinitely], the peaceful settlement of cross-Straits reunification through negotiations." Three days before the Taiwan elections, Premier Zhu Rongji angrily waved his finger, warning Taiwan voters: "Do not just act on impulse at this juncture which will decide the future course that China and Taiwan will follow... Otherwise I'm afraid you won't get another opportunity to regret" (BBC 2000).

Despite this pre-election rhetoric, the mainland reaction to Chen's victory was initially somewhat restrained. China, a foreign ministry spokesman noted, would "need time to listen to what they say and see what they do" – a refrain that would be used often during Chen's administration (Goldstein 2008: 16). However, this apparent openness was qualified by earlier warnings of dire consequences if Chen refused to recognize the one China principle.

Soon after Chen's victory, the 2000 presidential campaign in the United States went into high gear. There seemed to be little question that the Republican candidate, George Bush, and his party were taking stances favoring Taiwan. Although Bush consistently opposed independence, the charge that Clinton's policy in the Taiwan Strait had been "tilted toward Beijing" suggested that a Bush presidency might just correct the tilt. There was strong evidence to support one analyst's assessment that Chen Shui-bian "arguably enjoyed the best circumstances vis-à-vis the United States of any president of Taiwan since 1979" (Shih 2004: 4).

CHEN OVERPLAYS HIS HAND – CHINA AND THE UNITED STATES REACT, 2000–2003

When Chen Shui-bian became president he was faced with the task of managing the two major foreign policy challenges. On the one hand, he needed to re-establish the confidence of the United States by undoing the damage done by Lee Teng-hui's "surprises." On the other, he had to provide assurances to the mainland that, despite his own history and that of his party, he was prepared to approach cross-strait relations constructively. However, he was limited by political constraints and pressures at home. Chen was a minority president in terms of both his popular mandate and his party's position in the legislature. Despite obstruction by an opposition party stunned and embittered by its defeat, he would have to demonstrate to a sceptical public his own and his party's capacity to govern. His principal challenge in managing cross-strait relations would be to maintain the support of the DPP's base while winning over the support of those who favored a more open approach to the mainland. Most important in this latter grouping was the business community, which looked to the new administration to ensure a favorable climate in cross-strait relations. Chen was, thus, faced with performing a difficult political balancing act.

Appropriately, the first attempt to reconcile these pressures was in his inaugural speech of May 2000. Noting that "leaders on both sides possess enough wisdom and creativity to jointly deal with the question of a future 'one China,'" he offered a pledge – known as the

"5 noes" – that would become the foundation of his position on negotiations with the mainland.

> [A]s long as the CCP regime has no intention to use military force
> against Taiwan, I pledge that during my term in office, I will not declare
> independence, I will not change the national title, I will not push forth
> the inclusion of the so-called "state-to-state" description in the Consti-
> tution, and I will not promote a referendum to change the status quo
> in regards to the question of independence or unification. Furthermore,
> the abolition of the National Reunification Council or the National
> Reunification Guidelines will not be an issue. (Federation of American
> Scientists 2000)

Although the speech received a strongly favorable reaction from the
general public in Taiwan, political fissures were already appearing. The
KMT-controlled Legislative Yuan announced that it would establish
procedures for overseeing cross-strait relations. Among Chen's political
allies there were also signs of discord. It was clear he would have to
operate within the constraints imposed by an opposition KMT, the
more hardline independence members of his own party, and the more
general political necessity of maintaining his political standing within
the party as a whole. Given the opposition from the KMT and the
doubts of his own party, Chen had very little room to maneuver in his
mainland policy.

Still, Chen made three important conciliatory gestures in his first
year, besides the "five noes," to reach out to the mainland. The first
gesture, included in his New Year's address of 2001, was to acknowledge the "one China" nature of the ROC constitution and to propose
that "the two sides of the Taiwan Strait begin from economic and cultural integration, through which mutual trust may be gradually built
and a new framework laid out for eternal peace and political integration" (Liu 2001). Second, he declared later that year that he would

reverse attempts by the Lee administration to limit investment on the mainland and pursue a more open policy of "proactive opening and effective management" (Chen 2001). Finally, in January of 2001, Taiwan initiated what became known as the mini-three links of direct transportation between the offshore islands (Jinmen and Matsu) and the mainland (BBC 2001).

In essence, then, Chen's positive gestures regarding cross-strait relations were a pledge not to take provocative actions (thus easing Washington's concerns) and to loosen somewhat Taiwan's restrictions on economic relations with the mainland (a gesture to the business constituency during a time of economic recession). However, neither of these met conditions set by the mainland for a resumption of semi-official relations (acceptance of the one China principle) nor did they assuage Beijing's suspicions regarding Chen's untrustworthiness. Moreover, the impact of these relatively conciliatory gestures was further weakened by the mixed messages that emerged from the contentious political environment in Taiwan. One important example concerned the "1992 Consensus."

As we have seen, Beijing contended that the consensus was that each side would orally state their adherence to the one China principle. In June 2000, Chen remarked that he was willing to accept the 1992 Consensus, defined as "one China, with each side having its own interpretation" (a KMT formulation). Soon after, the chair of the Mainland Affairs Council, Tsai Ying-wen, clarified that this did not represent acceptance of the one China principle (Lo 2000). By the end of the summer, Chen was defining the consensus as an agreement to disagree while advocating something called "the spirit of 1992," which was characterized by "dialogue, exchanges, and shelving disputes" (Lin 2000).

Within a year, Chen's seemingly moderate stance on cross-strait relations was already showing signs of shifting. This was apparent during the run-up to the December 2001 legislative elections in his alignment with Beijing's *bête noire* of Taiwan independence,

Lee Teng-hui. During the campaign the issue of the island's separate identity played a major role, and the result was a stunning blow to the KMT, which lost nearly half its seats in the legislature, with the DPP taking first place and a fledgling independence-oriented party supported by Lee in fourth place. In his post-election comments, Chen spoke of the possibility of a DPP-led coalition (known as the Pan Green) in the legislative body.

Alignment with Taiwan identity versus the mainland had proven its political worth. Following the election, Chen moved further along that path. For example, during January 2002, there were discussions about changing the cover of the passport to identify it as coming from Taiwan; the Government Information Office removed the map of China from its logo; and a task force was established in the Foreign Ministry to explore the possibility of using "Taiwan" in the names on some of the official offices (Chu and Ko 2002). At the same time, Chen's administration was rejecting relatively moderate proposals from the mainland (discussed below) to facilitate direct transportation links that sought to work around the one China stalemate.

By the summer of 2002, it became clear that the trend toward an assertion of separate Taiwan identity, not conciliatory gestures, would become the dominant trend in Chen's management of cross-strait relations. On July 22, in his speech accepting the chair of the DPP, Chen declared that, if the mainland continued to reject his gestures, Taiwan would "go its own way." On August 3 he went even further. In a broadcast to the independence-oriented World Federation of Taiwanese Associations, he asserted that "there was a state on each side [of the strait]" and, claiming that Taiwan was "not somebody else's province," that Taiwan's future could be decided only by a popular referendum on the island (*Taiwan Communiqué* 2002).

Those around Chen claimed that Beijing's wooing of the island nation of Nauru away from Taiwan on the day he assumed the chair of the DPP was the cause of this outburst. However, it was likely that what figured more prominently in his calculations was frustration with the mainland's rejection of his initiatives, as well as a calculation of the political gains that might come from a stronger separatist stance. What is clear is that these statements, during the summer of 2002, marked the beginning of a new stage in the Chen administration. From this point on, conciliatory gestures to the mainland were few and far between. Rather, until he left office in 2008, Chen's presidency would be characterized by a nationalist agenda at home, provocation of the mainland across the strait, and the progressive alienation of the island's most important ally – the United States.

BEIJING – SCEPTICAL AT BEST

We have seen that Beijing's pledge that it would watch what the new administration would say and do before making a judgment was not without preconceived notions or qualifications. Even before Chen became president, the terms for reopening the non-official cross-strait talks, ended in 1999, were established: the new administration would have to recognize the one China principle (which Beijing claimed had been the case in 1992) and abandon the independence agenda of the DPP. These, of course, were concessions that Chen would not make.

Aware of this, Beijing changed tactics and sought to influence public opinion on Taiwan, especially in the business community. In January of 2001, PRC Vice Premier Qian Qichen used what one commentator called "standard united front terminology" when he proposed that China should "work together with Taiwan compatriots...that agree on one China...and unite with all the forces that can be united...to struggle against separatism." At the same time, Qian presented a broadened formulation of the "one China principle" that did not identify China as the PRC: "There is only one China, Taiwan and the Mainland are both a part of China and China's sovereignty is indivisible" (Brown 2001).

The principal target for Beijing's united front tactics was the business community. As we shall see in chapter 7, despite political tensions, these were years of strong growth in cross-strait economic relations. As had been the case since the 1990s, the business community was pressing for direct transportation links with the mainland, which would not only allow them to commute without having to go through Hong Kong but also to ship goods without going through a third destination. The mainland clearly saw an opportunity to gain support on Taiwan and suggested negotiations between business groups rather than the ARATS and the SEF. The Chen Shui-bian administration, despite its refusal to accept the mainland's prerequisite for such talks – the acceptance of the "one China principle" – countered that this was an appropriate topic for discussion between those "non-official organizations."

In its responses to the relatively moderate policies of Chen Shui-bian, during the first two years Beijing settled into a pattern of minimizing the positive gestures coming from the administration while characterizing the provocative actions as reflecting its true views. For example, after Chen's inaugural speech, Beijing held to its position that an acknowledgment of one China was the requisite for negotiations and depicted the speech as "evasive" and as proposing an agenda of "hidden independence" rather than "blatant independence" (Kwan 2000; Xinhuanet 2002).

Chen's statements in the summer of 2002 were the final confirmation of Beijing's assessment that he had not changed. A spokesman for the Taiwan Affairs Office of the State Council made this clear when he attacked Chen's "decision to cling to independence" and declared that his "separatist remarks will 'seriously sabotage relations between both sides of the Taiwan Straits and affect stability and peace in the Asia-Pacific region'" (Xing 2002).

In slightly more than two years after Chen's election, relations between his administration and the mainland had gone from stalemate to confrontation. During the same period, Taiwan's relations with the United States appeared to have entered a period of amity.

THE BUSH ADMINISTRATION: HOW TO LOSE A GUARDIAN ANGEL

In October of 2003, meeting with Chen Shui-bian during a stopover in New York, Therese Shaheen, the chairman and managing director of the American Institute in Taiwan, referred to President Bush as the Taiwan leader's "secret guardian angel." This might have been the case when Bush came to office in 2001. However, it was no longer so. Soon after she made the statement, Shaheen was dismissed, and two months later, in the presence of the Chinese premier, Bush criticized Chen publicly from the Oval Office with PRC Premier Wen Jiabao at his side. In less than three years Chen had succeeded in alienating not only the mainland but the island's most important supporter as well.

The incoming Republican administration of George W. Bush included a number of individuals more supportive of Taiwan and less concerned about the impact of that stance on Chinese sensibilities. For example, the incoming deputy secretary of defense, Paul Wolfowitz, along with the incoming deputy secretary of state, Richard Armitage, had signed an open letter in 1999 urging that the United States "make every effort to deter a form of Chinese intimidation...and declare unambiguously that the United States will come to Taiwan's defense in the event of an attack or blockade" (Tucker 2009: 255–62). This, of course, would be a significant change in the conditional language of the Taiwan Relations Act. However, it seemed to be echoed by the president himself on April 25, when, in a television interview, he asserted that the United States "would do whatever it took to help Taiwan defend itself." This not only further removed any ambiguity regarding the TRA but also went beyond the legislation by committing the United States to aid in Taiwan's defense. Clarification putting it in line with past American policy was almost immediate. However, the interview clearly suggested President Bush's inclinations regarding Taiwan (CNN 2001).

Moreover, in his early public statements the president went out of his way to acknowledge the relationship with Taiwan. For example, before his visit to Asia in February 2002 he "equated Taiwan with the Philippines, a formal U.S. ally." When he visited China, he referenced the TRA without paying the usual obeisance to the agreements with the mainland regarding Taiwan (Sutter 2002: 6). However, a more concrete sign of a tilt toward Taiwan by the Bush administration came a few days before that statement. Washington announced the largest Taiwan arms sales package in almost a decade. There were other signs of closer military cooperation. For example, members of the American military visited the island to observe the Taiwan military as well as to make needs assessments, and the president signed legislation that categorized Taiwan as a "non-Nato ally" for purposes of eligibility to purchase certain weapons systems. However, the most dramatic departure from tradition was the visit in March 2002 of Defense Minister Tang Yao-ming to the United States in order to attend a meeting of the US – Taiwan Business Council, where he encountered Deputy Secretary of Defense Paul Wolfowitz and Assistant Secretary of State James Kelley (Romberg 2003: 203–5).

Unprecedented courtesies were extended to official visitors to the United States from Taiwan. Unlike the tumult that accompanied Lee Teng-hui's visit to Cornell, on a number of occasions Chen Shui-bian was allowed to make transit stops in New York, Los Angeles, and Houston on his way to Central America. During these stops he met with local officials as well as members of Congress.

In 2002, Taiwan caucuses were formed in both the House and the Senate. One experienced Congress watcher and specialist on Chinese foreign policy concluded that, in the first two years, the Bush administration had not only carried out a "rebalancing of the U.S. stance in the U.S.–PRC–Taiwan triangular relationship in a direction favorable to Taiwan" but also had forged a policy that balanced American relations with both sides of the strait and with Congress (Sutter 2002: 10). Very soon, two developments upset that balance.

The first of these was the attack of September 11, 2001, which highlighted the importance of international cooperation in the "war against terrorism." Signs of apparent Chinese willingness to work with the United States were welcomed by the Bush administration. Moreover, with American attention focused on Iraq and Afghanistan (and, to a certain degree, on North Korean nuclear development), there would be little patience for tensions in the Taiwan Strait. The second was Chen Shui-bian's assertions during the summer of 2002 that Taiwan might "go its own way" and that there existed "one state on either side of the Strait." According to one commentator, the earlier, "positive atmosphere" in the relationship between Taipei and Washington "began to melt away," as the "Taiwan leadership … [appeared] more inclined to put personal political interests ahead of more strategic objectives and U.S. concerns" (Dumbaugh 2007: 7). Once again, as with Lee Teng-hui, Taiwan was a source of surprises for American policymakers.

In the next two years, and for his entire term of office after re-election in 2004, Chen would increasingly attune his policies not only to the assertion of Taiwan's sovereignty but to the establishment of a separate identity for the island and its political system. Washington would find itself both enmeshed once more in the cross-strait conflict and often in the uncomfortable position of responding to Chinese demands to restrain Chen lest serious consequences follow.

CHEN PLAYS THE IDENTITY CARD, 2002–2008

After 2002, Chen Shui-bian was a man with a mission, seeking to give substance to the claim that the administration he led was a sovereign international entity governing a nation with a distinctive political identity, uniting the population and defining its place in the world as separate from China. He pursued these goals not only because they were consistent with his own vision, but also because he believed that it was a politically expedient platform that would win the support of the majority of the people of the island for himself and the DPP. It was domestic politics that would drive an agenda that angered China, embroiled his government in one controversy after another with Washington, and ultimately failed to win the support of the Taiwan people (Yang 2008: 203–27).

As we have seen, by the end of Lee Teng-hui's administration, the study of the history and culture of Taiwan had become a legitimate academic subject and part of the secondary school curriculum. During the administration of Chen Shui-bian, this effort intensified. Its goal became not simply to promote a sense of Taiwan identity but also to diminish the Chinese identity that had been a part of KMT rule through what became known as a campaign of "de-Sinicization."

This campaign had many dimensions. There were efforts to place Taiwan history at the center of a secondary school curriculum, with most Chinese history included in the broader topic of world history. Historical exhibits were staged which highlighted not only the indigenous cultures of Taiwan but also the influences of early settlers such as the Dutch and the Spanish. Perhaps the most visible examples of this effort were the renaming of public facilities, publications, and institutions to diminish the association with China. Thus, the airport, which had been known as Chiang Kai-shek International Airport, was renamed after its location, as the Taoyuan International Airport. In addition, "China" was removed from the names of some state-run corporations, postage stamps were issued without the designation "Republic of China," and currency included images of Taiwan.

Chen's efforts to dissociate Taiwan from China were manifest in a number of more concrete ways. For example, in the fall of 2003, he pledged that, if he were re-elected in the following year, he would work to create a new constitution by 2006. He emphasized that the new document would deal strictly with issues of procedure (legislative size, relationship with the executive, etc.) and would not violate his inaugural pledges. However, it was clearly also a way of affirming the island's nationhood. As he told one audience, "Although Taiwan is an independent, sovereign country, it has been neglected by the international community...Promoting a new constitution and letting the people ratify it through referendum is a necessary act to transform Taiwan into a normal, great country" (Chang 2003).

For the next three years, the constitutional issue remained on the table. By 2007, with the constitutional pledge as yet unfulfilled, Chen bandied about the idea of "freezing" the present constitution and creating a constitution of the "second republic." The implication seemed to be that the links with the mainland of the "Republic of China" would be cut and a fresh "Taiwan" entity would be born. The severing of these links was also apparent in Chen's efforts basically to nullify the National Unification Guidelines and the National Unification Council. In January–February 2007, arguing that the organization and its mandate represented KMT policy and assuming that Taiwan was a part of China, Chen moved to "abolish" both. It was only with Washington's intervention that he agreed to a different formula whereby the council would "cease to function" and the guidelines would "cease to apply" – although the Chinese-language meaning was much closer to "terminate."[1]

This effort was soon overshadowed by the final, and for Beijing the most provocative, effort of the Chen administration to assert Taiwan's sovereignty and separate status from the mainland – securing membership in the United Nations. This was not the first time that application was made. It was Lee Teng-hui, under prodding from the DPP, who initiated the effort proposed by the island's allies in the UN using the title of "Republic of China." However, what made Chen's effort different was that the application was made under the name of "Taiwan" by means of a letter sent directly to the organization and signed "Chen Shui-bian, President, Taiwan." Chen was not shy about his motive in doing so, as he told a videoconference audience that "Taiwan will join the United Nations and Taiwan will become a new and independent, normal country." The letter was rejected, the stated reason being that, by admitting the People's Republic of China in 1971, the UN had accepted Taiwan's status as "part of China."

However, Chen was not done. He proposed that the issue of applying to the UN under the name of Taiwan be included as a referendum question in the presidential election of 2008. For the DPP, referenda were a means of getting the faithful to the polls and compensating for weakness in the legislature. As noted earlier, before coming to power, the DPP had proposed that any change in status vis-à-vis the mainland would have to be decided by a popular referendum. We have seen that Chen proposed this method in 2002 in regard to Taiwan's status and again in 2003 as part of a constitutional reform process.

Yet, besides alienating the United States and China, Chen failed to achieve his objectives. During the 2004 election, referenda proposed by the DPP regarding the mainland missile threat accompanied presidential balloting, but failed. In 2008, a referendum concerned with the application to the UN under the name of Taiwan also failed.

Indeed, speaking more generally, it is not clear that Chen was successful in recasting the identity of Taiwan at home and abroad. Domestically, after a strong showing in the 2001 elections to the legislature and victory in the 2004 presidential election, the DPP and Chen began steadily to lose support, suffering a crushing defeat in the 2008 legislative and presidential elections as well as various referenda. Moreover, public opinion surveys suggest that his policies had a limited impact on changing attitudes on the island. There were relatively small changes in how people identified themselves (Chinese, Taiwanese, or both) and a continued overwhelming preference for the status quo rather than independence (Mainland Affairs Council 2014a).

Moreover, Chen's administration did little to realize his campaign pledge of improving cross-strait relations. This is not to say that cross-strait interaction ceased altogether. As we shall see in chapter 7, trade and investment flourished and, by working through non-governmental organizations, charter flights were arranged that permitted Taiwan business people in China both to transport goods directly to the mainland and to commute to major Chinese cities on direct flights flown by airlines from both sides. Finally, as noted in the next section, after 2005 Beijing accelerated its united front or people-to-people diplomacy and established regular contacts with the KMT and its leadership, which paved the way for improved relations after the DPP's defeat in the 2008 presidential election.

Even more striking than the degeneration in relations with the mainland was the change in the relationship with the United States. After 2002, Chen Shui-bian managed both to squander the support of the Bush administration and to prompt it to engage in parallel efforts with the mainland to restrain him in an effort to secure stability in the strait area. China and the United States became the odd couple responding to Chen's policies.

THE UNITED STATES AND CHINA CONTAIN TAIWAN: 2002–2008

On December 9, 2003, during a joint press appearance with the visiting Chinese prime minister, Wen Jiabao, President Bush publicly rebuked Chen Shui-bian in an answer to a question regarding a referendum scheduled to accompany the presidential elections in March 2004. "We oppose any unilateral decision by either China or Taiwan to change the status quo. And the comments and actions made by the leader of Taiwan indicate that he may be willing to make decisions unilaterally to change the status quo, which we oppose" (US Department of State 2003). This was a remarkable gesture by a president who, almost three years earlier, had declared he was ready "to do what it took to defend Taiwan." What was equally remarkable was that the action that was seen as possibly changing the status quo seemed relatively innocuous – two referenda: one was a proposal that Taiwan develop an anti-missile system if the mainland refused to withdraw its missiles; the other asked whether the voters favored negotiations with the mainland to achieve "peace and stability."

However, the contents of the referenda were not the issue; rather it was the procedure itself that President Bush depicted as destabilizing because of Chinese concerns it would set a precedent for a future vote on independence. The United States was in the difficult position of justifying opposition to a democratic procedure because it might prompt the mainland to disturb the status quo by using force to solve the cross-strait dispute. There was another irony inherent in this incident. For more than fifty years the leadership in Beijing had protested American interference in the "domestic issue of Taiwan." Yet, now, the prime minister of China was standing next to the president of the United States as the latter did exactly that. In contrast to the many bitter conflicts over Taiwan in the past, the United States and China seemed to be cooperating. However, they were not so much cooperating as following parallel policies.

Chen Shui-bian's identity politics had presented both the United States and China with the challenge of devising policies to avoid a cross-strait conflict that could expand to become a Sino-American conflict. Neither Beijing nor Washington changed the fundamentals of its policy toward the area. However, both made important refinements in their approach to avert a crisis.

The most striking of these was Beijing's apparent reliance on the United States to contain Chen's provocations, as had been the case in December 2003 with President Bush's comments. Thus, Washington was critical when, in late 2004, the DPP ran a campaign, with strong overtones of Taiwan identity, promoting the idea of a new constitution to be ratified by a referendum. Chen's actions regarding the National Unification Council and National Unification Guidelines in early 2006 merited a visit from a special delegation, a rebuff from the State Department, and a warning from Deputy Secretary of State Robert Zoellick that "Independence means war. And that means American soldiers, sailors, airmen and marines" (*China Post* 2006).

It appears that the campaign to enter the UN under the name of "Taiwan" evoked the strongest concern from the mainland and the harshest reaction from Washington. The Chinese response to this effort reached an unprecedented crescendo. Chinese officials came to Washington to express concern that somehow Chen would find a way to promote his agenda, perhaps by declaring martial law. His last months in office were identified by the Chinese as a "period of great danger," during which Chen would use the referendum on the UN issue to establish Taiwan's *de jure* independence, an act likely to evoke a military response from the mainland.

The Bush administration took the Chinese warnings very seriously. Deputy Secretary of State John Negroponte echoed the Chinese position that the referendum was a step toward independence, while a member of the White House's National Security Council echoed one of President Clinton's "three noes" when he declared that "Taiwan or the ROC is not at this point a state in the international community" (Snyder 2007). Finally, in the fall of 2007, Thomas Christensen, deputy assistant secretary of state, outlined American policy toward Taiwan in a speech the title of which – "A Strong and Moderate Taiwan" – was a succinct statement of Bush's policy (Christensen 2007).

While Beijing clearly leveraged its relationship with Washington to restrain Taiwan, it did not soften its warnings to Taiwan itself regarding the consequences that might flow from Chen Shui-bian's policies. There were ample warnings of the consequences that might result should Chen continue these, amplifying the usual rhetoric about China's willingness to pay any price to prevent independence. However, as observers at the time noted, perhaps having learned its lesson from the reaction to Zhu Rongji's warnings before the 2000 election and responding to American policy, the Chinese leadership seemed to moderate its warning before the elections held during these years.

Besides looking to the United States and preparing for the worst, Hu Jintao, who succeeded Jiang Zemin as the head of the Communist Party in November 2002 and president in spring 2003, implemented a number of more positive policies intended not only to deter independence but to build a foundation for improved relations, despite the damage being done by Chen. In March 2005, at a meeting of the National People's Congress (NPC), Hu made his first major statement on the Taiwan issue in the form of a "four-point guideline on cross-straits relations under the new circumstances." The points were, first, "never sway in adhering to the one-China principle; Second, never give up efforts to seek peaceful reunification; Third, never change the principle of placing hope on the Taiwan people; and Fourth, never compromise in opposing the 'Taiwan independence' secessionist activities" (Xinhuanet 2005).

Soon after Hu's speech the NPC passed the Anti-Secession Law, which elaborated on these fundamentals by calling, on the one hand, for a broad range of exchanges and, on the other, for the achievement of "peaceful reunification through consultations and negotiations on an equal footing." However, it also legislated broadly defined conditions under which the use of force by the mainland would be justified:

> In the event that the "Taiwan independence" secessionist forces should act under any name or by any means to cause the fact of Taiwan's secession from China, or that major incidents entailing Taiwan's secession from China should occur, or that possibilities for a peaceful reunification should be completely exhausted, the state shall employ non-peaceful means and other necessary measures to protect China's sovereignty and territorial integrity. (National People's Congress 2005)

Having established a firm position on the essentials of China's mainland policy, Hu then took the surprising step in March and April of 2005 of inviting, separately, the chairmen of Taiwan's two major opposition parties (and running mates in the 2004 election as the "Pan-Blue Coalition"), Lien Chan (KMT) and James Soong (People First Party) to Beijing. It was an expansion of united front diplomacy, obviously intended to isolate Chen, to establish a channel of communication with other political forces in Taiwan, and to develop a favorable image for China after the Anti-Secession Law.

Given the greater size and political influence of the KMT in the Pan-Blue coalition, the visit of Lien Chan was the most important. It resulted in an agreement "to uphold the 'Consensus of '92', oppose 'Taiwan independence', pursue peace and stability in the Taiwan Strait, promote the development of cross-strait ties, and safeguard the interests of compatriots on both sides of the strait." It also specified "tasks" to realize these goals – among them to "promote a formal ending to the cross-strait state of hostilities [and] reach a peace accord" and to "establish a platform for periodic contacts" (BBC 2005). This relationship with the KMT would prove to be significant not only because the party played an important role in mediating and advocating for the business community. It was also an important channel of political communication across the strait that would help to lay a foundation for a new relationship when the KMT returned to power in 2008.

CONCLUSION: THE LESSONS OF THE CHEN SHUI-BIAN PRESIDENCY

The presidency of Chen Shui-bian was a challenging period in cross-strait relations. The mainland characterization of its last two years as a "period of high danger" seems appropriate. Chen was testing the tolerance of the mainland as he sought to construct a new national identity and to establish his government as a sovereign, independent entity known as Taiwan. Despite his insistent denials that this was his intention, his policies were coming dangerously close to Beijing's "red line," presenting the danger of a cross-strait conflict.

For the Bush administration, whose priorities were in other parts of the world, the last thing it wanted was a conflict in the Taiwan Strait triggered by the domestic political ambitions of Chen Shui-bian. And so, as the nature of Chen's vision for Taiwan became clearer, and it appeared obvious that he was beginning to exhaust not only Chinese patience but that of the Bush administration as well, Washington expressed its concerns directly and publicly. The implication of these statements was clear: if Chen continued along this provocative path, there would be consequences for Taiwan's relationship with its most important friend and protector. In short, both American and Chinese interests converged on the necessity of maintaining stability in the Taiwan Strait.

Both of these external reactions clearly had their impact on Taiwan's voters. During this, the second decade of Taiwan's democracy, there was surely a sense of liberation from the restrictions of KMT rule and a satisfaction in being able to reclaim some of the identity of Taiwan. However, by Chen's second term, election results suggested concern that his policies were going too far. Although it is difficult to glean from public opinion whether it was due to Chinese threats of force or American warnings or public reaction to economic difficulties and to increasing signs of Chen's corruption, his support eroded, the final blow being the KMT victory in 2008.

Although both China and the United States made their views known, it was ultimately Taiwan's newly enfranchised population that frustrated Chen's efforts and restored the uneasy calm in the strait. Thus domestic politics had been both the underlying cause of the "period of high tension" and the force that immediately ended it. In the next administration, Taiwan's domestic politics would once again be a driver of cross-strait relations – only this time it would be to frustrate efforts to reduce tensions.

6 | Satisfying Washington and Beijing

The decline in the popularity of the DPP administration which had been apparent during Chen Shui-bian's second term was confirmed by the legislative and presidential elections of January and March 2008. The "Pan-Blue Coalition" (the KMT and its allied smaller parties) increased the number of its seats in the legislature by more than 20 percent, occupying 75 percent of the 113 seats. The Pan-Green Coalition of the DPP and its allies was reduced to holding about 24 percent of the seats. In the presidential election, the KMT candidate, Ma Ying-jeou, received more than 58 percent of the popular vote. It was a stunning rebuke for the incumbent DPP.

There were many reasons for the defeat, most prominently charges of corruption (Chen was arrested soon after he left office) and the state of the economy. However, the deterioration in the state of Taiwan's relations with the United States and China was clearly an important factor. Ma had run as the anti-Chen candidate in regard to this issue. He pledged a policy of "no unification, no independence, and no use of force," while claiming that Taiwan would cease being a "troublemaker" and become a "peacemaker" instead. Finally, Ma endorsed the KMT position regarding the "1992 Consensus" (that it was an agreement that each side would state its own interpretation of "one China," with Taipei affirming that "one China" was the Republic of China) and proposed to open talks with the mainland on that basis.

A NEW RELATIONSHIP ACROSS THE STRAIT

In the first few months of his administration, Ma further elaborated on changes in Taiwan's cross-strait policies. He affirmed that negotiations would start with the mainland simply on the basis of the "1992 Consensus" regarding one China, while muting the identification of "one China" as the ROC, except when speaking to domestic audiences. He offered an agenda that would proceed on the basis of "first the easy and then the hard." Finally, he proposed that both sides skirt the sovereignty issue by pursuing an approach of "mutual non-denial" of the other's right to exist, thus avoiding the difficult question of the status of each side. In contrast to Chen's provocative efforts to enter the United Nations, Ma pledged that his administration would seek only "meaningful participation" in international agencies. Moreover, he tied success in overcoming international isolation to improved cross-strait relations and called for a "diplomatic truce" whereby neither side would "steal" the other's diplomatic allies.

Ma's election was greeted as an "historic opportunity" by the mainland. The formal meetings between the Straits Exchange Foundation and the Association for Relations Across the Taiwan Straits, which had been suspended in 1999, were resumed quickly under the rubric of the "1992 Consensus" (still defined differently by each side). Ma was anxious to deliver on his campaign promises of improving cross-strait relations, and the mainland was relieved to take advantage of the opportunity to resume dialogue with Taiwan.

It soon became clear that, in the mantra of "first the easy and then the hard," the "easy" consisted predominantly of economic questions. Between June 2008 and the end of 2014, twenty-one agreements were signed. However, as the process went on, it also became evident that, as the two sides were exhausting "easy" economic-related issues, it was proving difficult to move onto more challenging political or security issues. This was reflected by the fact that, in the first two years

(2008–9) of the Ma administration, fourteen agreements were signed, while in the entire first term (2008–2012) a total of fifteen of twenty-one were signed.

Discussing agreements touching on political questions has proven to be more problematic. These questions involve issues relating to the nature of the governments on each side of the strait – their sovereign status, the extent of their jurisdiction, and, most of all, their relation to each other. For example, suggestions by Ma of a peace treaty at the end of 2011, on the eve of the presidential elections, caused such a political firestorm on Taiwan that he eventually placed a series of conditions on its negotiation that assured its impossibility. Nonetheless, by the end of Ma's first term, the mainland was already calling for cross-strait talks to move beyond economic and cultural questions and into the "deep water" of political questions. However, with vociferous opposition from the DPP and increasingly unsupportive public opinion, Ma demurred. Political questions became relegated largely to discussions between non-government think tanks or, in the case of military matters, between retired officers. These discussions demonstrated the distance that separated the two sides on such questions.

However, in contrast to the *content* of cross-strait talks, which have moved very little beyond economic issues, there has been considerable development in the *levels* at which talks have been held. When dialogue was restored in 2008, the talks were restricted to the "non-government" organizations that had been established in the 1990s to manage cross-strait affairs (SEF and ARATS). Very soon these meetings were held at regular intervals, on either side of the strait alternatively. After the signature of the Economic Cooperation Framework Agreement in 2010, negotiations were being conducted by government officials at the vice-ministerial level on both sides. However, the most dramatic escalation in the level of interaction came in 2014, when the heads of the Mainland Affairs Council and the Taiwan Affairs Office exchanged visits. Although there had been Taiwan visitors meeting high-level

Chinese officials and vice versa, this was the first time that government officials at this level had met. The visits were a dramatic illustration of the considerable progress that has been made in building the quantity and quality of exchange mechanisms during Ma's administration and of the fact that cross-strait relations had entered their most stable period since 1949.

While Ma Ying-jeou might have achieved his stated goal of ceasing to be viewed as a "troublemaker" in the strait area, he has not done so well in regard to another objective – the enhancement of Taiwan's international profile. In this regard, Ma had put forth two objectives: to protect the handful of countries with which the ROC maintained official relations from switching over to relations with the mainland and to increase participation in international organizations.

The "diplomatic truce" has held through much of Ma's administration, with only The Gambia breaking relations with the ROC in 2013. However, it seemed clear that this had more to do with relations between the two than with the efforts of China to undermine the island's official relationships, and, in fact, Beijing did not establish relations with The Gambia after Banjul's break with Taipei. Taiwan's success in gaining participation in international activities has been more problematic.

Ma had made it clear from the beginning that how Taiwan was treated internationally would influence cross-strait relations. By choosing not to seek admission to the United Nations and announcing an approach of "flexible diplomacy," he suggested that Taiwan would avoid provoking the mainland by seeking "membership" organizations that required statehood (an effort not supported by the United States according to President Clinton's "three noes") but, rather, would seek to participate in such organizations. However, this has still required mainland acquiescence.

In 2002, an outbreak of Severe Acute Respiratory Syndrome (SARS), originating in China, became the object of coordination through the

World Health Organization, an agency of the United Nations. When physicians on Taiwan requested information on the disease, they were told to inquire of China, since the ROC was not a member. And, although a Taiwan representative was ultimately allowed to attend a conference concerned with the disease, its application for observer status in the WHO was denied. Before the Chen administration left office, it considered using admission as a topic for a referendum.

In 2009, Taiwan was finally allowed to take part in the World Health Assembly as an observer under the name of Chinese Taipei, the same title used at the Olympics. Although it was clear that the invitation would never have been issued without the assent of China, it was the first time since 1971 that Taiwan had participated in an international agency affiliated with the United Nations (deLisle 2009). In the fall of 2013, again with the assent of China, Taiwan participated as a "guest" (not an observer as requested) at the annual meeting of the International Civil Aviation Organization, again with the forbearance of China (US Department of State 2013b).

However, what is striking about cross-strait relations during the Ma administration is the fact that these improvements have been accompanied by increasing constraints imposed by public opinion on negotiation of not only the "hard" political questions but, as is discussed in the next section, even certain "easy" economic questions that touch on the interests of Taiwan society. Previously, we have seen how domestic political differences deeply influenced cross-strait policy during the administrations of Lee Teng-hui and Chen Shui-bian; their influence has been no less during the presidency of Ma Ying-jeou.

The fundamental element shaping the domestic political environment for Ma's mainland policy has been the bitter, partisan environment created by the cleavage between the KMT and the DPP. As we have seen, the origins of this rivalry go deep into the history of Taiwan and are sustained by elements of ethnic identity and mutual distrust. From the beginning of the Ma administration, the DPP continued to present itself as the political "other" to the KMT, opposing his policies at every turn.

On the subject of cross-strait relations, the DPP became the "party of no." As one observer has noted, the objections raised by the DPP to Ma's early initiatives went "across the board to include sovereignty, economic dependence, and even basic identity" (Romberg 2009). For example, even economic agreements with the mainland were pictured as trading sovereignty for economic gain and threatening the interests of Taiwan workers. For this reason, the DPP insisted that they be subjected to the closest scrutiny, and visiting officials from the mainland were met with demonstrations affirming the determination of Taiwan to protect its sovereign rights. However, most importantly, the DPP has refused to accept the suggestion that relations with the mainland could be anything but relations between two sovereign, independent entities – as in Chen Shui-bian's "one state on either side of the Strait." This was embodied in a 1999 party resolution which maintained that "Taiwan is a sovereign and independent country.... Taiwan, although named the Republic of China under its current constitution, is not subject to the jurisdiction of the People's Republic of China" (DPP 1999).

This resolution, which the DPP said superseded the 1991 call for an independent Taiwan by asserting the island's *de facto* independence by the DPP, has been the basis for the party's refusal to accept any version of one China such as the "1992 Consensus" in any form. Proceeding from these fundamental positions, the DPP has refused to recognize the basis on which talks were conducted with the mainland during the Ma administration while, at the same time, casting suspicion on the agreements reached.

When the DPP candidate for president, Tsai Ying-wen, was defeated in 2012, in part because of the lack of confidence regarding her ability to manage cross-strait relations, pressures developed within the party to reconsider its mainland policy. During the course of the campaign, Tsai did very little to change these fundamental positions. Rather, she continued the usual attacks on the KMT's policies as concessions toward the mainland and pledged to manage cross-strait relations responsibly while seeking a "Taiwan consensus" on their future management. However, after its defeat in the 2012 election, additional pressures built up in the DPP to reorient its mainland policy. Members of the party traveled to the mainland either in an individual capacity or as local politicians to initiate a dialogue. In addition, pressures built within the party to retreat from some of its fundamental positions.

For example, Frank Hsieh Chang-ting, who had served as premier in the Chen Shui-bian administration and was the defeated DPP candidate in the 2008 presidential election, promoted the idea of "constitutional consensus" to take the place of the 1992 Consensus when negotiating with the mainland. Based on the fact that both sides of the strait have constitutions that assume one China, this formula, rather than "one China, different interpretations," was thought to be an acceptable mantra under which to conduct negotiations. Although Hsieh traveled to the mainland and met with officials concerned with cross-strait relations, his proposal gained little traction either there or within the DPP (Wang 2012).

Despite studies and discussions, the DPP dithered over this and other proposals regarding cross-strait policy throughout 2014. By the summer of that year, when the party held its national congress under the chairmanship of Tsai Ying-wen, a proposal to "freeze" the clause was never considered. Indeed, Tsai canceled the discussion of the independence clause, claiming that independence was already a "natural ingredient" for the youth of Taiwan. With the local elections coming up in November, Tsai seemed to calculate that the DPP could do well without changing its approach to cross-strait relations and, moreover, that it was safest to avoid anything that might alienate the party base (Hung 2014).

The result of the local elections in 2014 was a stunning defeat for the KMT. The DPP was victorious in five of six mayoralty elections in Taiwan's major cities, with the party's total national vote in all elections going down to the village level at about 47 percent (*China Post* 2014). Of course, these were local elections that turned on a number of issues. However, it was both inevitable that there would be speculation regarding the impact that Ma's cross-strait policies might have had on the KMT defeat and likely that new pressures might develop for the DPP to formulate a more positive approach to the mainland that could take it into the 2016 presidential elections.

Still, while cross-strait issues were not directly contested in the campaign in November 2014, data concerning public opinion did suggest considerable popular ambivalence toward Ma and his mainland policy in the run-up to the elections. To begin with, the most commonly cited survey on cross-strait relations reveals an overwhelming popular preference for the maintenance of the status quo (either indefinitely, with a decision later, with later independence, or with later unification) rather than either immediate independence or unification with the mainland.[1] In all, 85.5 percent of those polled expressed a preference for the current status quo, with 59.5 percent favoring indefinitely or a decision later as options for the future. However, if given a binary choice between independence and unity, one 2013 poll by a Taiwan television station found that 78 percent favored independence and 18 percent favored unification (TVBS Poll Center 2013).

Viewed from a slightly different angle, but seen by some as supportive of the public's preference for maintenance of the present distance from China, is the issue of self-identification, as there has been a dramatic change in how residents of the island identify themselves since Ma became president. In 2008, 48.4 percent identified themselves as Taiwanese, 3.8 percent as Chinese, and 32.7 percent as both Chinese and Taiwanese. In 2014, the figures were 60.6, 3.5, and 32.5 percent respectively (Election Study Center 2015b).

When Ma Ying-jeou ran for the presidency, he sought to appeal to this mood among the voters of Taiwan by presenting himself as the candidate who would bring stability back to relations with China and who would not change the status of the island vis-à-vis the mainland. However, despite the fact that he has won two elections, Ma's favorability ratings have suffered throughout his time in office, and at the end of 2014 nearly 70 percent of those surveyed said they did not trust him, while only 11.5 percent were satisfied with his performance.[2]

This has been the result of a combination of elements, including the bitterly partisan nature of Taiwan politics, natural disasters, and factors relating to the island's relationship with the world economy. However, there are indications in the polling data suggesting Ma's declining ratings are the result of public ambivalence concerning his cross-strait policies. For example, public support seems to be strongest for the institutional framework that has been established for cross-strait communication. Thus, the creation of representative offices, the exchange of official visits, or even permitting tourists from the mainland are all viewed favorably in surveys. However, there are greater concerns when it comes to economic agreements (Mainland Affairs Council 2014b). Here the data suggest public concern regarding the impact of cross-strait agreements on the island's economy. For example, one survey reported that more than half of the respondents felt that the mainland was benefiting more than Taiwan from economic agreements, while others found concerns of a similar nature regarding the consequences of cross-strait agreements on employment opportunities.

Lurking behind these concerns is some evidence of public distrust of Ma and his administration when it comes to the specific question of managing or negotiating cross-strait affairs. More than a majority of those polled expressed doubt that his government would protect Taiwan's interests in agreements or its sovereignty, with those dissatisfied outnumbering those satisfied by almost two to one.

It seems the very same domestic politics favouring the maintenance of the status quo that stymied Chen Shui-bian's efforts to enhance Taiwan's sovereignty have similarly hampered Ma's efforts to improve and institutionalize cross-strait relations. The preponderance of opinion among the Taiwanese population has had the paradoxical effect of frustrating policies that are perceived as either challenging Beijing on the issue of Taiwan's autonomy or endangering that autonomy by enhancing ties with the mainland. Suspicion of the mainland in regard to both its coercive abilities and its drive for future unification appears to deter independence sentiment even as it slows rapprochement.

When he ran for the presidency, Ma Ying-jeou successfully captured this mood with his platform of neither independence nor unification, only to run afoul of the mood later in his administration, when he was perceived by the public as upsetting the balance with a tilt, if not directly toward unification, at least toward the policies that would inevitably lead to unification. The overwhelming preference for the status quo reflected in public opinion surveys appears to define the mood of Taiwan's electorate as democratization in Taiwan enters its third decade. Of course, this public mood challenges the DPP as the 2016 election approaches. The lessons of Chen Shui-bian's administration and the 2008 election are that, to be successful, the DPP must demonstrate that it can satisfy Washington's priority of maintaining peace and stability in the strait area while managing the relationship with the mainland. To secure victory, this, of course, requires that the party and its presumptive presidential candidate in 2016, Tsai Ying-wen, craft a cross-strait policy satisfactory to both governments as well as to the party's supporters and a sufficient number of non-party voters.

The formulation of a policy acceptable to the mainland is likely the necessary first step in achieving this balance. As we have seen, throughout 2014 the DPP struggled with a formulation and achieved little success. Another effort has been launched since the November elections. However, the DPP must not only satisfy Beijing that it is a different party from the one that governed from 2000 to 2008 – as Tsai said in her announcement that she would run for president – it must satisfy mainland prerequisites for the continuation of the relationship established in the Ma administration that are, in their present form, incompatible with DPP doctrine. Thus, the mainland has its own policy challenge as the Ma Ying-jeou administration comes to an end. It must find a way either to provide new momentum to the relationship that has languished in the last years under Ma and, especially, should the DPP regain power in 2016, deal with a party which it distrusts deeply and which refuses both to recognize the validity of the "1992 Consensus" and to foreswear Taiwanese independence as the basis for cross-strait negotiations.

BEIJING AND MA YING-JEOU: PATIENCE . . . BUT FOR HOW LONG?

Even before Ma Ying-jeou's inauguration as president in May of 2008, it was clear that the mainland was looking forward to a new era in cross-strait relations. The month before, in the highest-level meeting between a Chinese official and an elected official from Taiwan since 1949, President Hu Jintao of China met with Vincent Siew, the vice-president-elect of the ROC. After the meeting, Hu noted that, while, in the past, relations with Taiwan had "suffered twists and turns for reasons known to all," the meeting with Siew had "inspired us to think deep about cross-strait economic exchanges and cooperation under the new circumstances" (*Taipei Times* 2008).

Although Ma's inaugural address expressed the KMT's version of the "1992 Consensus," which declared the ROC to be the "one China," it was still good enough for Beijing to agree to reopen the talks between the SEF and the ARATS which had been suspended in 1999. A meeting in June 2008 settled some important issues regarding air travel, and six months later another meeting was the occasion for signing the "three links" across the strait – air, sea, and postal. In the same brief period of time, the head of the SEF traveled to the mainland, followed by a reciprocal visit by the head of the ARATS, who met with President Ma and Minister Lai of the Mainland Affairs Council – the highest-level visitor from the mainland to date and the first to visit with President Ma.

These fresh initiatives were followed by an important new programmatic statement regarding the conduct of cross-strait relations delivered by Hu Jintao in December 2008. In this speech, Hu laid out six principles that would guide mainland policy in the future.

The first of these was that both sides of the strait should maintain the "one China principle" that "There is only one China in the world. China's sovereignty and territorial integrity brooks no partition" and independence should be opposed. Secondly, both sides should promote extensive economic cooperation "to lay a more solid material foundation and provide a greater economic impetus for the peaceful development of cross-Straits relations." The third principle was that, while the value of Taiwan's culture would be recognized, "both sides should jointly inherit and promote the exquisite traditions of Chinese culture." The fourth was to promote exchanges to enhance mutual understanding, coupled with the hope that the DPP would "realize a clear understanding of the tenor of the times, cease carrying out its 'Taiwan independence' secessionist activities and stop running counter to the common wish of the entire nation." If its members did so, they were assured a "positive response" from the mainland.

The final two principles contained the most important concrete initiatives directed at Ma's government. The fifth touched upon the delicate issue of Taiwan's external relations. It recognized the controversial nature of the issue and pledged that "further consultations" could be held on "people-to-people economic and cultural interactions with other countries" and added:

> Regarding the issue of Taiwan's participation in the activities of international organizations, fair and reasonable arrangements can be effected through pragmatic consultation between the two sides, provided that this does not create a situation of "two Chinas" or "one China, one Taiwan." Settling the Taiwan question and realizing the complete reunification of the country is an internal affair of China and is not subject to interference by any foreign forces.

Finally, the statement acknowledged that "the two sides may make pragmatic explorations in their political relations under the special circumstances where the country has not yet been reunified [and] may at an opportune time engage and exchange with each other on military issues and explore mechanisms for mutual military trust." It ended with an appeal that, "on the basis of the one-China principle, we should formally end the state of hostility across the Straits through consultation, reach a peace agreement, and build a framework for the peaceful development of cross-Straits relations" (Taiwan Affairs Office 2008).

This speech both established the inflexible bottom line of the mainland's approach to Taiwan after the election of Ma and suggested those areas where it might be willing to explore mutual agreements. In regard to the former, the obvious firm positions were the insistence on acceptance of the "one China principle" and the opposition to independence as well as to interference by the United States. In contrast, the potential for considerable flexibility was contained in the phrase "the special circumstances where the country has not yet been reunified." This was an implicit acceptance of a period of indeterminate length during which the two sides would not be united and yet engage in negotiations that touched not only on economic issues but also on questions related to Taiwan's participation in international affairs or cross-strait security, with confidence-building mechanisms and a peace agreement both specified as desirable outcomes.

That these principles were a response to Ma's more patient approach of the easy economic issues first and the hard political issues later seemed confirmed when negotiations began for the Economic Cooperation Framework Agreement in 2010. At that time, Beijing made it clear that it was willing to make significant economic concessions to overcome suspicions on Taiwan and to keep the momentum of agreements going. Beijing set aside its political agenda for the moment and focused on building good will on the island through tourism and trade – the policy of "winning the hearts of the Taiwan people."

Still, toward the end of Ma's first term there were occasional comments made by mainland officials which suggested impatience regarding the failure to move on to the more "difficult" issues of political and security relations. However, apparently as a result of the sharp public reaction to Ma's discussion of a peace agreement at the end of 2011, Beijing eased up on the demand, perhaps concerned that pressing on the question would work to the benefit of the DPP during the upcoming presidential election.

In 2008, there was no real question as to which side Beijing favored in the presidential contest, although its statements lacked the harshness of Premier Zhu Rongji's warning in 2000. Leaders on the mainland warned that, unless the opposition DPP was willing to accept the basis for the cross-strait talks (the "1992 Consensus" on one China) and abandon its position on independence (neither of which it did), the positive momentum established over the past four years would be lost. As one might expect, China celebrated Ma's election for a second term.

Ma's re-election in January 2012 was followed in November by the Eighteenth Congress of the Communist Party and the political transition from Hu Jintao to Xi Jinping. In his last speech as general secretary of the party, Hu strongly affirmed the mainland's commitment to peaceful reunification and called on both sides to "uphold the common stand of opposing Taiwan independence and of following the 1992 Consensus." Finally, with the Taiwan election over, Hu expressed the "hope that the two sides will jointly explore cross-straits political relations and make reasonable arrangements for them under the special condition that the country is yet to be reunified," specifically mentioning confidence-building mechanisms and a peace agreement as possible agenda items (Xinhuanet 2012).

It should have been no surprise that the mainland would press for political discussions once Ma was elected. Underlying the agreements on economic and cultural issues was Beijing's assumption that these would create the foundation of mutual trust necessary for political negotiations – the "easy" would pave the way for the "hard." In the period after 2012, at a time when the pace of agreements was slowing, Beijing pressed harder for discussions of political questions.

The most dramatic example of this came in October of 2013, when China's new leader, Xi Jinping, met with Taiwan's former vice-president and commented on the agenda of cross-strait relations, noting: "the issue of political disagreements that exist between the two sides must reach a final resolution, step by step, and these issues cannot be passed on from generation to generation" (Reuters 2013). Xi's remarks signaled a new direction in mainland cross-strait policy. The importance of addressing political questions soon became a prominent theme in mainland statements, with the minister in charge of mainland affairs cautioning that, although "some political differences can be shelved temporarily, it is impossible to avoid them totally or for a long time" and that "paying attention only to economics and not politics is not sustainable." Moreover, additional evidence that Beijing was becoming impatient with the slowed pace and substance of cross-strait negotiations could be found in the linkages being made by some important leaders between the period of "peaceful development" and "reunification," seemingly making clear that the latter was by no means an open-ended time frame (Romberg 2014b).

As we have seen above, because of the sensitivity of political talks, Ma Ying-jeou's administration resisted these initiatives, and, eventually, political discussions were relegated to "track two" discussions between academic organizations, think tanks or organizations of retired military officials. These meetings were often attended by officials from both sides, as well as, in the case of Taiwan, members of different political parties – including the DPP. However, rather than resolving issues, such meetings simply illustrated the difficulty of dealing with political issues, especially the most fundamental of all – the status of each side and the relationship between the two. The ambiguous "1992 Consensus" and mutual non-denial were merely ways of avoiding the issues. The mainland's concept of "one China" simply could not be reconciled with the ROC's claim of its own sovereign status. And, until the question of the nature of and the relationship between the entities negotiating on each side was resolved, little could be accomplished in the political realm.

Still, however frustrated the mainland might have become during Ma Ying-jeou's second term, it shared with his administration the importance of maintaining the momentum of cross-strait relations. This was likely one of the reasons why there was an exchange of visits by cabinet-level officials handling cross-strait relations on both sides in 2013–14, noted earlier. It was a significant enhancement of the relationship likely intended as a way to keep up the momentum of cross-strait relations.

Efforts to create a more mainland-friendly environment on Taiwan went beyond relations with the Kuomintang authorities. There has been an ever widening variety of mainland delegations traveling to Taiwan for the purposes of making business contacts, visiting with local officials or civil society groups, or attending non-governmental meetings of various kinds. At the individual level, mainland students are attending Taiwan colleges and universities and mainland tourists visit the island.

Indeed, two events in 2014 caused the mainland to intensify these grassroots efforts. The first was the so-called Sunflower Movement, a student-led movement organized in opposition to the ratification of the trade in services agreement with the mainland (discussed in chapter 7) in the spring of that year. The second was the stunning defeat of the KMT in the November elections. In the wake of the student movement, in May of 2014, Xi Jinping mused that the mainland might have been talking to the wrong people in Taiwan and that it was necessary to focus more on the grassroots and younger people. Not coincidentally, when Zhang Zhijun, head of the Taiwan Affairs Office of the State Council, visited Taiwan a month later, he was said to have "expressed his hopes for more communication with local people and to hear their advice on cross-Strait cooperation." This was substantiated in the itinerary of his trip, which highlighted visits with local people in the pro-DPP area in the south of Taiwan. Similar themes were present after the November 2014 elections (KMT 2015).

However, it is not clear how successful these policies have been. As we saw in the last section, surveys of the Taiwanese people show very little enthusiasm for unification, identification with Taiwan is strong, and there is considerable skepticism regarding the benefits, to the island, of economic agreements. These attitudes are even stronger among younger people than other age groups. Moreover, since Ma took office, the percentage of those polled who described the mainland attitude toward the Taiwan people as "hostile" fell below 40 percent only once, and never when it came to the mainland attitude toward the Taiwan government (Mainland Affairs Council 2014a).

In this blitz of people-to-people diplomacy, there is one very prominent organization that the mainland refuses to deal with – the DPP. Although it has welcomed party members, current and former office holders, and party-sponsored think tank delegations in their individual capacities, it has rejected any suggestion of dealing with the DPP as an organization unless it abandons its independence platform and accepts the 1992 Consensus (or some version of the "one China framework").

Amid rumors after the elections of 2014 that the DPP was beginning to reconsider its mainland policy, Beijing showed little inclination to be flexible. Commentators continued to demand that the party abandon its advocacy of independence and accept the "1992 Consensus," which was depicted as the "anchor" of cross-strait relations. As noted earlier, there have been attempts within the DPP to adopt statements that might offer an alternative perspective (Frank Hsieh's "constitutional consensus") or a "freeze" of the independence platform. Moreover, in the period after the November 2014 election, Tsai Yingwen was clearly aware of the need to address the issue of the party's mainland policy, and discussions within the party have begun. However, with less than a year to the presidential election, there has been no adjustment in DPP policy that approaches the mainland's criteria for negotiations and no sign of Beijing's flexibility on its requirements for dealing with the party.

There can be little doubt that Beijing has become impatient with the pace and substance of cross-strait relations during the later years of the Ma administration. It has attempted to maintain momentum by finding areas ranging from transit stops for mainland travelers in Taiwan to concluding two-year-old discussions regarding the establishment of SEF and ARATS offices on each side of the strait. Facing the reality of the situation in Taiwan, the Chinese leadership seems to have decided to adopt a more patient, long-term strategy that seeks to maintain this momentum, to change perceptions of China at the societal level in Taiwan, and to avoid actions that might generate support for the DPP in the presidential and legislative elections of January 2016.

However, there are also indications in mainland commentaries of a growing impatience with the lack of response to its gestures, perceived as generous, and there is an increasing awareness of the restraints imposed upon the Ma administration by the sharply divided public opinion on the island. If this mood remains unchanged or, even more challenging, if Beijing is faced with a DPP administration after 2016, maintenance of a future status quo may rely on the patience of China's leaders.

THE UNITED STATES: RELIEF AND CONTINUED COMMITMENT

There was an almost palpable sigh of relief in Washington when Ma became president. The Bush administration, even President Bush himself, had tired of the provocative policies of Chen Shui-bian. It was anticipated that Ma Ying-jeou's pledge to be a "peacemaker and not a troublemaker" would lower the temperature in cross-strait relations and, by doing so, simultaneously lower the levels of Washington's exasperation with the government in Taipei, and therefore reduce direct American involvement in the region.

However, the easing of cross-strait tensions did not change the longstanding differences between the United States and China on the Taiwan issue. Nor did it reduce growing American concerns regarding China's rise and growing self-confidence. The fact that Chen Shui-bian had managed to violate the conceptions of an acceptable status quo held by both China and the United States allowed for a parallelism in policy that gave the illusion of American and Chinese cooperation and, along with Chen's equivocation over arms sales (see chapter 8), diminished the salience of the Taiwan issue as a point of conflict in US–China relations. So, although the historical Sino-American differences over Taiwan had been obscured, they had by no means been resolved. Thus, as the new administration took office in Taiwan, the first question that many American commentators asked was how the outreach of the Ma government to the mainland could affect the cross-strait policy of the United States.

Some analysts highlighted China's rise, the limits to American power, and the Ma administration's willingness to negotiate. It was proposed that Washington wash its hands of involvement in cross-strait issues and simply abandon Taiwan, leaving it up to the two sides to resolve matters between themselves. As one commentator put it, the island was simply one of the "vestigial positions that we [the United States] can no longer afford to support" (Rothkopf 2010). In a more imaginative and optimistic variant of this view, another analyst saw Ma's new policies toward the mainland as possibly resulting in the "Finlandization" of the island and, thus, solving the problems caused in Sino-American relations by the Taiwan issue (Gilley 2010).

For other commentators, the shift in domestic politics on Taiwan made possible a more active American support of the island as a response to a rising China. Chen Shui-bian's provocations of the mainland had frequently muted Taiwan's support in Washington. However, for strong supporters of Taiwan, the island's current, less provocative policy toward the mainland appeared to provide an opening to argue for a firmer American commitment. Some depicted the island as occupying a crucial geostrategic position in the face of a rising China, while for others it was the shared values of democracy that required expanded relations with Taiwan. This school of thought argued for strengthening ties in the military or economic areas, with the most extreme versions of this view arguing for an abandonment of the "one China" policy and the recognition of Taiwan's status as a sovereign state.

Of course, the two extreme prescriptions of abandoning Taiwan for the relationship with China or leaning to the side of Taiwan at the expense of relations with the mainland have long provenance in debates over Washington's policy. However, since the 1970s, the mainstream policy has been one that seeks to balance the two approaches by means of a commitment to an American-defined status quo (neither Taiwan independence nor mainland coercion) and addressing the cross-strait controversy through peaceful negotiation.

During the initial period of Taiwan's new relationship with the mainland there were some rumblings of concern in Washington that the Ma administration might drift into the orbit of the mainland, with consequences for American security interests in the area. Some wondered whether Washington would really like to see a peaceful settlement. However, as the more positive relationship evolved, the message from Washington was one of encouragement of the changed tone and the agreements that emerged in cross-strait relations.

There was no dramatic realignment of Taiwan policy. However, overall, the American interest in the area was defined as it had been for decades – "peace and stability." The new administration greeted the reopening of cross-strait negotiations and the agreements reached by the two sides. As the relationship evolved during the two terms of the Obama and Ma administrations, it became clear that Washington saw Ma's governance as consistent with American interests in a peaceful, stable strait area and provided important support as well as encouragement.

Of course, the most concrete, and arguably most important, expression of American support remains in the military realm – the security relationship with arms sales at its core, which will be discussed in greater length in chapter 8. The consultation and cooperation between the two nations' militaries continues and remains exceptionally close, considering that Taiwan is neither a state nor a treaty ally. Taipei and Washington have agreed on the importance of arms sales as both a significant contribution to the defense of Taiwan and a factor demonstrating American support and providing the island with the confidence to pursue cross-strait negotiations.

As of the end of 2014, congressional notifications of arms sales to Taiwan by the Obama administration totaled $12 billion. Of course, the relationship has also had its tensions. The United States still decides what it will and will not sell, and during the Obama administration, as we shall see below, Taiwan has been disappointed in its attempts to secure advanced fighter planes and assistance on conventional submarine manufacturing. On the other hand, Washington has been impatient with the Ma administration's failure to demonstrate its commitment to its own defense by meeting its commitment to a defense share of 3 percent of the national budget.

The renewal of large-scale arms sales by the Obama administration was bitterly criticized by China. Amid the usual charges that the United States was in violation of past agreements, as well as Chinese sovereignty, was using Taiwan as an instrument to prevent China's rise, was upsetting the positive trend in cross-strait relations, and was giving confidence to independence elements in Taiwan, the mainland threatened retaliation that ranged from boycotts of defense contractors to discontinuance of Sino-American military exchanges (Romberg 2010). From the perspective of the mainland, arms sales remain the symbol of American refusal to truly accept Beijing's definition of one China and to abide by past agreements. It remains a non-negotiable issue.

A second, less tangible (but to Taiwan equally important) area of American support during the Ma administration has related to participation in international organizations. As we have seen, one of the principles set down during the Clinton administration was that the United States would not support Taiwan's participation in any organization requiring statehood. Washington has not supported efforts by Taipei to gain admission to the United Nations. However, when Ma Ying-jeou sought to "participate meaningfully" in international agencies that did not require statehood, the Obama administration followed the precedent of earlier administrations and provided support.

CONCLUSION

It is clear that the Taiwanese–American relationship during the Ma administration has been restored to a condition that is arguably the best that it has been since democratization began in the 1990s. Ma's pledge not to be a troublemaker has created an environment quite different from that of the administrations of Chen Shui-bian or Lee Teng-hui. The support described above, as well as intangibles such as visits of relatively high-level American office holders, the granting of visa-free entry for Taiwanese passport holders, the poorly disguised preference for Ma during election time, and the flexibility shown during his transit stops are all indicative of this.

However, none of this has served to ease mainland concerns about the role of the United States in cross-strait relations. When Chen Shui-bian's presidency came to an end, the existence of parallel interests in maintaining the status quo, which had eased Sino-American tensions over Taiwan, also ended. As the mainland sought to woo the government in Taipei as well as the population of the island toward commitment to closer ties with the mainland, the behavior of the United States once more has come to be seen as an impediment to this goal. In meetings with the Chinese, American interlocutors were continually reminded that Taiwan is a "core interest" of China, and strong protests regarding American arms sales (even while only gently chiding Taiwan) reflect a return to mainland suspicions regarding US motives in respect to the island.

And so, after more than six decades, the United States is still in the middle of the cross-strait conflict, seeking to maintain its relations with both sides locked in an irreconcilable relationship while, at the same time, serving as the key factor in deterring an armed conflict between them.

The development of cross-strait economic relations over the past twenty years stands in sharp contrast to the volatile and difficult inter-governmental relationship discussed earlier. It seems that political issues have not intruded on a relationship driven by converging economic interests. Except for brief periods, economic ties have maintained a steady upward trajectory of growth and increasing complexity. However, the line between politics and economics must be drawn carefully. Although each of these two dimensions has had its own trajectory and its own actors, neither side has lost sight of the impact that the economics could have on the achievement of its political aspirations.

This chapter traces the evolution of cross-strait economic relations from the end of the twentieth century until the present. It begins by focusing on the changing nature and size of the relationship as well as its importance to each side. It then explores the manner in which economic and political issues have influenced the policies of each side and relations across the strait. The chapter concludes with a discussion of the future impact of economic issues on cross-strait relations.

SOME STATISTICS

Two-way trade across the Taiwan Strait has grown dramatically since it was sanctioned by Taipei in the late 1980s.[1] When Lee Teng-hui was elected president in 1996, it amounted to US$3.7 billion (US$18 billion). It grew to $10.6 billion ($30.5 billion) when Chen Shui-bian was elected in 2000 and then to $98.3 billion ($129.2 billion) when Ma Ying-jeou was elected in 2008. In 2013 it was at $124.4 billion ($197.2 billion).

Cross-strait investment has shown a similar pattern of growth – at least in respect to Taiwan's investment on the mainland (mainland investment in Taiwan was only allowed beginning in 2009). In 2001, the cumulative number of investments on the mainland that had been approved by the Ministry of Economic Affairs in Taipei since 1990 was 24,160 (50,838 were reported by the mainland), though by 2012 the number of approvals had grown to 40,208 (88,001 according to mainland figures). In comparison, since 2009 there have been a mere 342 cases of approved mainland investments in Taiwan.

PATTERNS OF GROWTH AND COMPOSITION

To get a complete picture of the nature of cross-strait economic relations, these numbers have to be further parsed.[2] The extent of dependence on cross-strait trade for each side, as well as the balance of trade and its composition, reveals much about the nature of the economic relationship.

Trade with China and Hong Kong represented 14.7 percent of Taiwan's total in 1996, rising to 15.2 percent in 2000. However, by the end of Chen Shui-bian's presidency in 2008, the figure had risen to 27.7 percent, with the mainland becoming Taiwan's leading export market (38.9 percent) but trailing Japan in terms of imports (13.6 percent to 19.3 percent). In 2013, China and Hong Kong retained their status as lead export market, at 40 percent, but remained very slightly behind Japan in imports from the mainland, at 16 percent to 15.7 percent.[3]

The percentages cited above show that Taiwan has enjoyed a favorable balance of trade with the mainland. In 1996, the ratio of its mainland exports to imports was approximately 7:1. This figure dropped to about 5:1 in 2000 and to 3:1 in 2008 as well as in 2012. However, it remains an important part of the island's overall trade, contributing nearly 50 percent of its global favorable balance. However, the composition of that favorable balance has shifted since the early 1990s. In 1993, eight of the top ten exports to China were footwear and various items related to textile manufacture. In 1996 and 2000, this number dropped to three, and by 2008 and 2012 they were no longer among the top eight. What grew enormously during this period and eventually came to dominate Taiwan's exports to China were electrical machinery, mechanical appliances or parts, plastics, iron and steel, and optical equipment/testing machines. In 1996, these five represented 53 percent of exports to the mainland. By 2008, they constituted 75 percent, and, if iron and steel were omitted from the list, the remaining four represented 94 percent (in 2012 the four, 73 percent). They also represented 42 percent of Taiwan's global exports in these goods, which, in turn, constituted over 50 percent of the island's global exports.

The pattern of mainland exports to Taiwan was strikingly similar. In 1996, the four items identified above plus fuels represented 48 percent of the island's imports from China, rising to the mid-60s in 2008 and 2012. Significantly, between 2000 and 2012, the two categories of machinery nearly doubled and came to represent 84 percent.

The importance of the mainland to Taiwan's economy is also apparent in the island's outward investment. Like trade, this has grown steadily since the end of the twentieth century, with the mainland taking up an estimated 70 to 80 percent of all overseas investment from Taiwan. In 2012, more than 75 percent of these investments have been in the manufacturing sector, followed by the service sector, with growing investment in banking and finance as well as the retail/wholesale trade. Taiwan is the largest foreign investor in China, but its importance to the mainland economy as a whole is only 15 to 20 percent of total inward investment. Moreover, mainland investment in

Taiwan has been permitted only since 2009 and has fewer than 400 approvals by the Taiwan government.

ANALYZING THE NUMBERS: THE DEVELOPMENT OF ECONOMIC RELATIONS

The first conclusion to be drawn from these statistics is the apparent dependent relationship that Taiwan has with the mainland. China absorbs the bulk of the island's foreign investment. It is its major export market in goods that constitute the most important of Taiwan's exports and, finally, it is the source of a large, favorable trade balance, which, in turn, makes possible a positive global balance for the island's economy. Secondly, these trade statistics provide an insight into the evolution of Taiwan enterprises on the mainland, from the manufacture of simple consumer goods to the production of sophisticated electronics.

In the late 1980s and early 1990s, Taiwan's small and medium enterprises, which produced labor-intensive goods for export, were facing severe challenges. At home, land and labor costs were increasing, and a growing environmental movement was complicating production. The appreciation of the Taiwan dollar made Taiwan products more expensive and neo-protectionist sentiment in the island's traditional export markets was growing. When restrictions on investment abroad were reduced in the late 1980s, Taiwan business people – primarily in labor-intensive "sunset industries" – began to look abroad to areas with lower production costs, more flexible environmental regulations, and larger import quotas. Given the cultural affinity and geographic proximity, China was a natural choice.

During this initial period of Taiwanese investment on the mainland, companies manufacturing non-durable consumer goods such as toys, plastic household goods, and shoes dominated. In time, the larger, upstream suppliers for these manufacturers (for example, Formosa

Plastics) and companies producing for the mainland market (such as the food industry) moved to the mainland.

By 1995, Taiwanese firms had entered the computer age. Between 1994 and 2004 the proportion of investment on the mainland devoted to information technology (IT) more than doubled, reaching 56 percent. Taiwan companies were producing more than half of global production in components such as monitors and mice as well as 54 percent of laptop computers. Much of this production was on the mainland. By 2002, 55 percent of desktop and 40 percent of laptop production by Taiwanese companies was on the mainland. At the same time, the investors in China became larger, publicly listed firms. This was reflected in the size of the average investment in China, which went from US$735,000 to US$2.78 million between 1991 and 1995 (Chin 1997: 173). Moreover, the major investment sites shifted from the Pearl River Delta to the Yangtze River Delta – particularly in the Shanghai area and in Jiangsu, in cities such as Suzhou, Kunshan, and Wujiang.

The close relationship between mainland investment and the composition of Taiwan's exports to the mainland reflects the third characteristic of cross-strait economic relations as they have evolved since the 1990s. This is the "investment induced effect," where Taiwan's exports to the mainland are orders from Taiwan-invested companies for mainland operations (Luo 1999: 135; Tung 2002). These materials are then processed by less expensive Chinese labor and shipped back to Taiwan or to world markets. For example, between 1995 and 1998, 44 percent of raw materials and 49 percent of parts and semi-finished goods were exported to Taiwan enterprises.[4] Moreover, mainland imports to Taiwan reflected a somewhat similar pattern. In certain higher-value-added industries, components or semi-processed goods produced in China were returned to Taiwan for finishing and export abroad. Thus, after the 1990s, cross-strait trade and investment on the mainland were intimately linked, with cross-strait trade being, in reality, inter- or intra-firm trade among Taiwanese companies.

However, not all production materials for Taiwan enterprises were imported from the home island. As production moved to the mainland, so did suppliers. Very early in the investment process, a "clustering" effect became apparent, with upstream suppliers concentrating their mainland investments in areas convenient to their manufacturing customers. By 1998, 23 percent of parts and semi-finished goods were procured from other Taiwanese firms by manufacturers on the mainland. Yet few of the products made in China were Taiwanese brands. In the 1970s Taiwan's manufacturing enterprises were original equipment manufacturers. This meant the manufacturing of products according to specifications determined by a foreign firm that were then sold under that firm's brand name. Similarly, Taiwanese companies initially produced IT equipment under the name of foreign companies such as HP or Sony.

Finally, and most ironic of all, was the fact that the growing number of IT products made by Taiwanese enterprises on the mainland, using components from Taiwanese companies, were not only being sold under foreign brand names but being classified as Chinese exports. As such, Taiwanese industries played a vital role in China's export-driven economic development, with about half of the mainland's exports produced by them.

Today, cross-strait economic relations continue in the manner described above. As the statistics cited earlier illustrate, the relationship remains asymmetrical in regard to both investment and trade, and manufacturing remains the predominant form of Taiwanese enterprise on the mainland. However, the Taiwanese manufacturers who flocked to the mainland in the 1990s have also evolved in response to global developments. As demand for computers declined, companies have moved into areas such as the manufacture of data-storage equipment to meet the needs of the growing cloud computing markets. They are also making smartphones and tablets. Perhaps the most famous of these companies is Hon Hai (Foxconn), which evolved from a manufacturer of electrical connectors in the 1990s to become the employer of more than half a million Chinese workers in several locations in China and the principal manufacturer of Apple iPhones. Hon Hai also produces for Sony (Japan), Nokia (Finland), and Dell (US).

Moreover, Taiwan's IT manufacturing on the mainland has gone beyond original equipment manufacturers. Companies such as the laptop manufacturer Quanta, which claims to build a third of the world's laptops, also provides original design manufacture, in which the producing firm contributes to the design of a product sold under the name of a foreign firm. Finally, a few Taiwanese companies manufacture under their own brand names as well. This is the case with Acer computers and HTC smartphones.

Today, the location of investments as well as the clustering effect continues the earlier trends. The destination for Taiwanese investment is still in the Yangtze River Delta, as companies that supply upstream components follow their buyers. Although mainland firms are increasingly joining these supply chains, the linkages among Taiwanese firms remain central, especially for certain more complex components, for reasons ranging from simple reliability to past relationships.

The increasing clustering of Taiwanese firms has had a major impact on local communities. For example, Kunshan, in the Suzhou/Shanghai area, is a Taiwanese community, with restaurants and schools catering to those who have resettled from the island. Indeed, in many areas where Taiwanese industries are clustered, there are associations of business people organized to represent the interests of these industries vis-à-vis local officials.

When viewed from a national perspective, Taiwanese industries are an important part of the Chinese drive to build a modern economy through export-oriented industrialization. However, at the local level, where economic success is directly tied to the career prospects of local officials, Taiwanese investment makes a vital contribution. One study of its impact on the Jiangsu area found that Taiwan had provided 53,

80 and 90 percent of foreign direct investment in Suzhou, Kunshan, and Wujiang, respectively, while another study calculated that, at one point, Taiwan-invested enterprises provided Kunshan with more than 60 percent of tax income and 90 percent of foreign trade (Chen et al. 2010: 214; Lee 2006: 11).

However, inter- or intra-firm trade still takes place across the Taiwan Strait. This can take many forms. In most cases the headquarters of a Taiwanese firm remains on the island, although some companies are incorporated offshore or funnel investments from offshore (this accounts for the large amounts of investment that come to China from the British Virgin Islands). In addition, a company might take orders from international customers on Taiwan or maintain its research and development facilities there to utilize local talent or protect intellectual property rights. Finally, in some cases, more complicated production or machinery will be manufactured in Taiwan (integrated circuits, parts for smartphone cameras, etc.) and shipped to the mainland as exports.

This growth and complexity of cross-strait economic relations during the past two decades is remarkable given the fact that, for most of this time, relations touching on non-economic matters were either non-existent or stalemated. Until the election of Ma Ying-jeou in 2008, the architecture of cross-strait relations was overwhelmingly the result of unilateral policies adopted by each side rather than a structure resulting from bilateral negotiations. This does not mean that political pressures and priorities have been absent in the economic policies adopted. In establishing its own policies, and later in negotiating with the other side, politics was very much part of each side's economic policies.

A POLITICIZED TRADE REGIME, 1990–2008

The mainland has not been shy about asserting that these contacts served an important political purpose. As Yang Shangkun, then China's president, put it in December of 1990, the purpose of economic exchanges was to "*use* business to exploit politics and utilize the public to urge the official [*yi shang wei zheng yi min bi guan*]. We should lead cross-Strait exchanges in the direction to facilitate unification of the motherland and the four modernizations" (Tung 2002: 2).

Economics would create a united front from below, as China put its "hope in the Taiwan people" rather than in its government. For example, during the second half of the 1990s, as cross-strait political relations deteriorated, mainland statements sought to mobilize this constituency by emphasizing the importance of improved relations for Taiwan's economic health. This message intensified (especially in respect to the economic benefits of direct shipping) as the island's economy braced for an unprecedented decline in the early years of Chen Shui-bian's administration and was present throughout much of that time even as Beijing kept up the attacks on his "independence policies."

In pursuit of these political goals, the mainland created an exceptionally open and supportive environment for Taiwanese investors, who were granted many of the advantages given international investors. Foreign trade legislation specified that, despite the island's status as a "region" of China, they were granted status as "foreign investors." Although Beijing has refused to discuss an investment protection treaty with Taipei because it would suggest that relations were international in nature, other legislation was intended to ease the concerns of Taiwanese investors. These regulations covered matters ranging from tax status to the schooling of children, to the protection of property, to the enjoyment of equal rights with Chinese firms with respect to the purchase of materials and utilities. Provinces and cities on the mainland set aside special zones for investment by Taiwanese enterprises and provided added incentives such as tax exemptions, duty-free imports, land-use rights, and so forth – some of which were granted without the approval of the central government.

Moreover, while enhancing the political strength of the KMT was clearly the principal motivation for the interparty dialogue begun in 2005, the relationship between the two parties also worked to the advantage of Taiwan's economic community. For example, an agreement reached at the fourth round of meetings in 2007 resulted in mainland pledges to take actions regarding intellectual property rights and investment protection. Finally, the vital economic role of Taiwan-invested enterprises was recognized by the formation of local business associations intended to advocate for, and protect the interests of, Taiwan businesses on the mainland (Chien and Zhao 2010: 255).

In sum, political objectives were a very important consideration in the making of China's cross-strait economic policy in the late twentieth and early twenty-first century. During years when the political atmosphere was tense, economic ties grew at rates that have not been seen since. For China's leaders, frustrated in their ability to make progress toward unification, economic ties could serve as something of a surrogate for promoting that goal, even as they contributed to the mainland's export-driven economic development.

Beijing's motives did not go unnoticed on Taiwan. During the administrations of both Lee Teng-hui and Chen Shui-bian, the burgeoning economic relationship was viewed with concern. Taiwan's efforts during these years were directed toward creating a trade regime that would frustrate the mainland's efforts to use economics to promote political goals and maintain the status quo of separation, while at the same time meeting domestic pressures from the business community to expand the trade regime. At a time when Taiwan was progressively liberalizing its trade vis-à-vis the rest of the world, it was seeking to restrict interaction with the mainland. These efforts had little success. Economic relations flourished amid political tensions.

Soon after Chiang Ching-kuo permitted limited cross-strait trade and investment, the government became alarmed over its impact on the island's economy and its ability to resist mainland pressures. As early as 1990, there was official concern over "mainland fever" and frequent warnings of the dangers of a relatively small amount of Taiwanese trade and investment with the mainland. In 1992, the Lee Teng-hui administration sought to stem the tide by promulgating an act that established broad governmental powers to regulate "dealings between the people of the Taiwan area and the mainland area." Pursuant to the program of "avoiding haste and being patient" launched by Lee after the 1996 election, a number of more specific regulations to limit trade and investment were put into effect that included specifying permissible imports from, and types of investments in, the mainland.

These efforts had little success. Although, in a few prominent cases, Lee was able to divert investment away from the mainland, the restrictions had a very limited effect. The impact of democratization increased the political influence of the business constituency and Taiwanese companies found ways to circumvent the laws. Either the regulations were simply ignored or, more commonly, dummy companies would be established in Hong Kong, the United States, or even the British Cayman Islands for the purpose of evading the scrutiny of the Taiwan government.

With the approach of the 2000 presidential elections, the DPP sought to cultivate business interests by publishing a White Paper that pledged that the party would "adopt an open attitude to economic and trade development" on the mainland.[5] In his New Year's message in December 2000, Chen Shui-bian signaled a readiness to expand economic relations in the hope that it could aid in "building a basis for a new framework of permanent peace and political integration." By early 2001, with the Taiwan economy in recession, Chen organized the Economic Development Advisory Conference. During the summer, a recommendation was made that Lee Teng-hui's "no haste, be patient" policy be replaced by one of "active opening, effective management."

However, by 2002, policy toward mainland trade was changing again. As we saw in chapter 5, by the middle of that year there were signs of change in Chen's position on cross-strait relations, marked by his declaration in the summer that "Taiwan and China are each one country on each side of the strait." His more moderate policy had not been well received by many in his own party or by the independence-oriented Taiwan Solidarity Union. With the mainland ignoring his more conciliatory position, and the danger that his political support within the green camp was eroding as the 2004 election approached, Chen hardened his position on cross-strait relations. By March 2006, the approach had developed further and a policy entitled "Active Management and Effective Opening in Cross Strait Economic and Trade Relations" was adopted. This "new concept" was, of course, the reverse of the earlier slogan, which had called for "active opening," and sent a signal that policy was shifting from one of encouragement of mainland trade to one of restraint. Indeed, for the remainder of his time in office, Chen resisted the scenario he had suggested in 2000 of economic integration leading to political reconciliation (Brown 2006).

Still, if one looks solely at the statistics of the Chen years, the economic relationship appears to have had a life of its own and was able to grow through evasion and the use of existing mechanisms despite the unsupportive cross-strait political environment. There had been no negotiated agreement since 1993 and no significant meeting of representatives from both sides since 1998. The economic environment described above was simply a structure that was made up of unilateral policies formulated by each side to achieve political objectives. Any efforts to change this pattern through mutual agreement proved almost impossible, as illustrated by the fate of two issues: direct transportation across the strait and each side's policies within the World Trade Organization (WTO).

The business community and shipping interests had demanded direct cross-strait transportation links.[6] With the exception of the "mini-links," the existing regulations mandated that travel to and from the mainland had to be indirect. Business people returning to the island were required to make one stop (usually in Hong Kong or Macao) and either change planes or stay on a plane then given a different flight number before continuing to Taiwan. Similarly, goods shipped to and from Taiwan were required to make a stop at an intermediate destination before continuing. A research institute on Taiwan estimated that direct links would not only save valuable time for those business people virtually commuting between the island and the mainland but also yield significant cost savings of 16 percent for goods and as much as 27 percent for individuals.

Under intense political pressure, Chen had pledged to implement talks on direct links by the end of 2004. However, this offer was an empty gesture as far as the mainland was concerned. When talks were broken off in 1999, Beijing had pledged not to return to the table for talks between the SEF and the ARATS until Taiwan accepted the "one China principle," and its proposal for talks among private groups was rejected by Chen. Finally, in 2006, compromise was reached. Taipei, having passed legislation that allowed industry associations to be authorized by the government to negotiate, permitted a delegation headed by the Taipei Airlines Association with government representatives present to negotiate with the Chinese. The result of these negotiations was an agreement on cross-strait charter flights.

This agreement was the exception that proved the rule regarding the politicization of economic relations. It was arrived at because direct links was an issue consistent with the political and economic interests of both sides. For Chen, it garnered support from an increasingly impatient business community, and for the Chinese leadership it served to maintain ties during a difficult time in cross-strait relations. It was the most important of the very few negotiations that took place between the years 2000 and 2008. But it did not, as some predicted, set a pattern for further agreements.

The salience of competing political goals in shaping economic relations was also apparent in the conditions under which both sides entered the WTO in 2001, a successor to the General Agreement on Tariffs and Trade (GATT). The stated purpose of the GATT was to arrive at "mutually advantageous arrangements directed to the substantial reduction of tariffs and other barriers to trade and to the elimination of discriminatory treatment in international commerce." Having agreed to begin the process of achieving major tariff reductions in the import of foreign goods, the signatories established two fundamental principles that would continue to guide its successor organization, the WTO: most-favored nation and national treatment. The former assured that concessions granted to one member would be available to all, while the latter prevented discrimination (such as taxation) against imported goods in the home market.[7]

There was considerable optimism regarding the economic potential if both sides of the strait attained WTO membership. The existing size and quality of the economic relationship seemed to provide precisely the kind of environment that could thrive in the WTO trade regime if it conformed to the organization's rules. However, this was not the case. Taiwan maintained restrictions on 38 percent of agricultural imports and 24 percent of product imports from the mainland, regulations regarding mainlander travel to Taiwan, severe limitations on mainland investments in the island's service industries, and restrictions on types of investment permitted on the mainland. All of these were clear violations of the core principles of the WTO.

For the mainland, the difficulties did not revolve around concerns regarding increased trade and investment. Overall, expansion rather than restriction served the PRC's economic and political goals. But, with Taiwan, extension of these privileges touched on the bedrock of the mainland's cross-strait policy – that economic relations with Taiwan were a domestic matter between parts of one China. Compliance with the rules of the international organization would appear to validate the island's claim to sovereignty and permit it to negotiate with the mainland on an equal basis. This was precisely the problem as far as the mainland was concerned. Throughout the admissions process, the mainland tried to link Taiwan's membership to that of China. These efforts failed. Taiwan was admitted as the "government" of the "Separate Customs Territory of Taiwan, Penghu, Kinmen, and Matsu." However, once they were both admitted, the mainland refused to treat Taiwan as having equal status in regard to organizational matters. In that sense, WTO membership did little to improve cross-strait relations.

In sum, for nearly two decades economic ties thrived and grew steadily despite the volatile state of cross-strait relations. In this sense, economic relations developed on a track separate from that of politics. Still, the influence of politics was not absent. Each side sought to shape economic policy to conform to its political priorities, but with varying degrees of success. In these years, an essential part of the global technology economy enjoyed vigorous growth under a trade regime that was the result of self-serving rules unilaterally established in the absence of any significant negotiation by two governments seeking competing political ends. After the election of Ma Ying-jeou in 2008, the unilateral nature of the rules of cross-strait trade changed; however, the impact of politics did not.

THE POLITICAL LIMITS OF ECONOMIC NEGOTIATIONS, 2008–2014

As we have seen, Ma Ying-jeou campaigned for the presidency on a platform that included a pledge to deepen cross-strait relations. Economics was, of course, a logical first step to take. Despite the tensions of the Chen Shui-bian years, a complex fabric of relations had been woven between the two sides. Moreover, as noted earlier, the Kuomintang had served as a *de facto* representative for Taiwanese business interests on the mainland in the last years of the Chen administration.

It was not long after the election that there were signs that the unilateral structuring of cross-strait economic relations would be supplemented by negotiated agreements. It was obvious that when, in his inaugural address in May of 2008, Ma called for talks with the mainland on economic issues based on the "1992 Consensus," he was knocking on an open door. It was not long before the two semi-official bodies formed in the 1990s, the ARATS and the SEF, met for the first time since 1998 and began to address a backlog of issues. On June 13, 2008, less than a month after Ma's inauguration, the heads of the two organizations met in Beijing and agreed to direct, scheduled flights as well as to mainland tourism to Taiwan (BBC 2008).

Although both sides would continue to make unilateral adjustments to the conditions of trade, this was the beginning of a new phase in economic relations. By the end of 2014, twenty-three agreements or memorandums of understanding had been signed. Their subjects ranged from the protection of intellectual property rights, to nuclear safety, to securities and futures regulation. In addition, informal communication between government agencies became common. If problems arose, it was not uncommon for officials to pick up the phone to call their counterparts on the other side.

However, clearly the most significant agreement in the emerging, bilateral cross-strait economic architecture was the signing in June of 2010 of the Economic Cooperation Framework Agreement (ECFA). One commentator described its significance as "the most important agreement between these two political rivals since the end of the Chinese Civil War in 1949." This agreement identified four goals for the future cross-strait economic regime:

1 gradually reducing or eliminating tariff and non-tariff barriers to trade in a substantial majority of goods between the two parties;

2 gradually reducing or eliminating restrictions on a large number of sectors in trade in services between the two parties;

3 providing investment protection and promoting two-way investment;

4 promoting trade and investment facilitation and industry exchanges and cooperation.

Moreover, in two annexes, it identified a limited number of goods and services where restrictions or tariffs would be relaxed six months after signing – an "early harvest" – pending expansion in future agreements. Finally, an administrative body, the Cross-Strait Economic Cooperation Committee, was established with representatives from both sides in order to oversee the implementation of ECFA and the drafting of subsequent agreements. ECFA was intended eventually to replace the many economic agreements with one overarching structure, then to go on to eliminate the limitations on compliance with the WTO rules and negotiate a free trade agreement (Hsieh 2011).

However, for Taiwan there was a new, compelling motivation to achieve a completed agreement with the mainland: the proliferation of free trade agreements in East Asia with China that have excluded Taiwan. The danger for Taiwan was obvious. If no agreement were reached with China, its economic relations with the mainland would be subject to restrictions and tariffs from which others would be exempt, thus increasing the cost of its manufacturing on the mainland. This danger was highlighted in late 2014, when South Korea and China announced the intention to complete a free trade agreement. The challenge was obvious. According to one report, 77 percent of Korean exports are similar to those of Taiwan, and some of these goods will now be entering China under the free trade agreement, placing Taiwanese goods at a disadvantage (Chyan 2014).

Advancing the ECFA to completion is important in a second respect. Taiwan has free trade agreements with only a small number of its trading partners. It has been assumed in Taipei that, before it could seek free trade agreements with other Asian nations, it would have to secure an agreement with, as well as a positive signal from, China.

Thus, the specific goods and services agreements that would result from the ECFA would not only benefit cross-strait economic relations but would also be a necessary step in assuring Taiwan's inclusion in a rapidly changing East Asian economy.

Still, from the beginning, the DPP has opposed Ma's program of liberalizing mainland trade. Despite the fact that Chen Shui-bian had presided over a huge expansion of Taiwan's trade with the mainland, the Ma administration's efforts came under immediate attack from the DPP opposition. Thus, even though ECFA established only a frame-work for further economic liberalization, it was immediately criticized as endangering Taiwan's sovereignty and economy. In the period between the beginning of negotiations and the signature of the agree-ment in June 2010, the opposition party expressed its opposition. Although the agreement was ratified by the KMT-dominated legisla-ture in August after a stormy debate, there would be specific agree-ments yet to come, and the DPP had laid down their marker.

Mainland negotiators were obviously aware of the sensitive nature of the agreement and sought to allay concerns on Taiwan. After all, its goal was not purely economic; it was also intended to create a public mood more favorable to reunification. Thus, even before the agreement was signed, Beijing signaled that it would take into consideration Tai-wan's concerns regarding the possible impact on the island's economy – especially the agricultural sector. Yet the political currents in Taiwan were not so easily calmed. Trade liberalization and internationalization are volatile political issues in most political systems because of the threats posed to a range of domestic economic interests. Given its past protectionist trade and investment regime, as well as the political weight of the agricultural sector, Taiwan is not an exception. However, as the initial reaction to the new administration's mainland policy demonstrated, these difficulties were intensified when economic self-interest was combined with suspicion regarding the political motives that lay behind mainland economic policies toward the island.

This became dramatically obvious when it came to a consideration of the cross-strait trade in services agreement signed pursuant to the ECFA agreement in June of 2013. This agreement opened eighty mainland sectors to Taiwan investment and sixty-four sectors on the island to mainland investment (Xinhuanet 2013). The DPP opposition immediately attacked it on two fronts. The first was that it would result in a flood of mainland workers entering the job market, to the detriment of local workers. The second was that the agreement would further undermine Taiwan's sovereignty (DPP 2014).

When the agreement came to the legislature for review, the DPP's seizure of the podium made consideration difficult and the later occupation of the legislature by student demonstrators made it impossible. These spring 2014 demonstrations, which became known as the Sunflower Movement, soon expanded and appeared to be driven by concerns regarding both the threat to Taiwan's sovereignty and the impact on the economy (Romberg 2014b). By the end of the year, the agreement had not cleared the legislature, and it seemed that it would only be contemplated once a legislative basis for considering agreements with the mainland was passed. The negotiation and passage of agreements under ECFA suggests that even the "easy" economic questions are becoming "hard," and it is not difficult to see why.

The cross-strait agreements signed early in the Ma administration did more to institutionalize and facilitate existing trade patterns than to change or expand them. However, the ECFA signed in 2010 envisioned a trade regime that went not only beyond the existing, spontaneously developed pattern of trade but also beyond the elements of the WTO regime then in effect. It indicated a new orientation in economic relations. The agreement had sparked opposition from economic interests fearing the deleterious impact of the agreements, and from society as a whole, which, as we have seen in the previous chapter, is concerned with the political consequences of cross-strait agreements.

Finally, the new stage in Taiwan's economic relationship with the mainland and its quest for inclusion in emerging trade regimes has introduced another complicating factor – the response of the United States. During the administration of Chen Shui-bian, Washington's attention to cross-strait relations had been focused on the avoidance of conflict. When the tone of relations changed during the Ma administration, greater attention was paid to the economic aspects of the relationship and to the possible consequences of the developing economic architecture in Asia.

THE UNITED STATES AND CROSS-STRAIT ECONOMICS

There were, of course, American interests at stake in the growth of the cross-strait economic relationship. China and Taiwan were, as we have seen, vital links in the Asian production chains that were developing at this time. The Taiwan factories on the mainland produced computers and peripherals for American companies such as Dell and Hewlett Packard that, along with other consumer goods, were staples of exports to the United States. Taiwan chip foundries were a key source of supply for American companies, and there was active cooperation between Silicon Valley and science parks in Taiwan.

In contrast to the voices raised on Taiwan during the Chen administration expressing concerns about trade with the mainland, American business interests there appeared to greet or even encourage the expansion of the economic links. For example, the American Chamber of Commerce in Taiwan decried the efforts of Chen's administration to limit the development of economic relations and pressed for the establishment of direct cross-strait links. The response of the US government was to welcome any evidence of peaceful contacts between Taiwan and the mainland and to promote their further development. Although the election of Ma Ying-jeou and the sudden warming in economic relations alarmed some, who saw Taiwan sliding into the mainland's grip, the reaction of the State Department was to express the "hope those relations will continue to expand and develop" (US Department of State 2010).

When Taiwan sought to leverage its agreements with the mainland and the United States into entrance into the evolving free trade structures in Asia, American involvement became more direct. Since its accession to the WTO in 2002, Taiwan had been seeking a free trade agreement with the United States. However, apart from the difficult relations between Taipei and Washington during the administration of Chen Shui-bian, there were other questions, ranging from restrictive economic regulations to copyright issues, that made such an agreement extremely unlikely. Moreover, during both the Chen and the Ma administrations, the importation of American meat products into Taiwan was an especially rancorous issue. As noted by the then American envoy to Taiwan, William Stanton, difficulties in settling an agreement over the import of beef due to the presence of additives had become "the symbolic embodiment of Taiwan's protected markets" and a bar to a free trade agreement with the United States (Shih 2012).

The complications with the United States were symptomatic of Taiwan's apparent political difficulty in making the reforms that would be necessary for adherence to a free trade regime. This situation was a serious one, not only because of the importance of improved trade relations with Taiwan's only major ally in the world, but also because it touched upon the Ma administration's efforts to end its marginalization and join the Trans-Pacific Partnership (TPP).

As part of the "pivot" (or "rebalance") to Asia, the Obama administration in 2011 joined what was then the Trans-Pacific Economic Partnership Agreement, which had been created by Singapore, Chile, Brunei, and New Zealand. Between then and the present, the organization has expanded to twelve nations (including Japan), changed its name to the Trans-Pacific Partnership, and become the platform for negotiating a comprehensive trade agreement that would go beyond the WTO in its scope to regulate everything from intellectual property to trade in services (Meltzer 2011). Since the United States, but not China, is a founding member, and since joining does not require statehood, membership of the TPP presents a more likely amelioration of Taiwan's isolation than membership of ASEAN-based organizations, including the proposed Regional Comprehensive Economic Partnership, of which China is a member. Although the launching of the TPP is by no means assured, it remains a more promising option for Taiwan, and, with the unusual support of the DPP, Ma has proposed that it prepare to join by 2020.

Thus, successfully negotiating a free trade agreement with the United States is also important because it is seen as easing entrance into the TPP. Still, this will not be an easy task. The meat controversy with the United States demonstrated not only the apparent inability of the government, on account of internal divisions, to follow through on agreements but also the strong influence of interest groups. Although it is not unique to Taiwan, public opinion on the island is not generally supportive of lowering trade and other barriers. Moreover, should China decide to seek membership in a future TPP, it is likely that, as was the case with the WTO, it would be given the courtesy of joining before Taiwan. Since Chinese compliance with the organization's rules is likely to take longer than Ma Ying-jeou's target of 2020, this will extend Taiwan's time outside the new trading system (Bush 2014). Finally, China could put obstacles in the way of Taiwan's participation. Beijing could pressure other countries involved in planning for TPP to exclude Taiwan or, as it has indicated in the past, insist on a successful completion of ECFA before it considers the island's participation in regional economic organizations.

To sum up, what began with the migration of a few small-business people to the mainland, to continue to manufacture inexpensive toys, and then resulted in a major global center for the manufacture of communications and technology equipment, has become a link in a global production chain, which may possibly morph into part of a new Asian trade regime. This evolutionary process has resulted in the United States, on account of its certain influence over the form Asia's future trade regime will take, being drawn into cross-strait economic relations while also providing additional leverage to China.

CONCLUSION: ECONOMICS AND CROSS-STRAIT RELATIONS

We have seen that politics is very much a part of the economic relationship between China and Taiwan, and has been so since the 1990s. Although the future trajectory of the relationship is likely to be shaped by changing global production chains and multinational trade agreements, its defining nature remains a relationship in which one side seeks to use economic means to achieve political unity while the other seeks to achieve economic goals and resist such unity. How economic relations may change this delicate – and potentially dangerous – balance has been the subject of considerable speculation, with some seeing economic ties as ameliorating the conflict and others seeing them as a potent weapon that favors the mainland as the dominant economic actor in the relationship.[8] The former view is based on the assumption that the economic cooperation of international actors, and the benefits that can be derived from such cooperation, can ameliorate differences and progressively condition a peaceful relationship by increasing the economic price of conflict and the economic benefits of collaboration. The latter view is almost its mirror image and argues that increased economic contacts where one side is dominant can be an instrument with which to coerce the other side by exploiting its dependency.

There seems to be little evidence to support the proposition that the development of cross-strait economic relations during the past twenty-five years has served to lessen the political differences across the strait. As we have seen in chapter 5, the public opinion trends and the policies of those in office have shown little movement from a commitment to Taiwan's sovereignty. In fact, during much of this period (the administrations of Lee and Chen), the growth in the economic relationship occurred at times when political relations were strained. Most recently, despite the dramatic changes in cross-strait economic relations, attempts to deepen the economic relationship during the administration of Ma Ying-jeou have been resisted by constituencies committed to Taiwan's sovereignty and still distrustful of the mainland.

Of course, there are business interests that might look more favorably on concessions intended to ease relations with the mainland. However, while there has been some evidence of those within the business community opposing policies that might provoke the mainland, only few have actually advocated making political concessions.

On the mainland, official statements as well as policy (the "early harvest" concessions on trade, for example) suggest that there is a belief in Beijing that economic policy will change political attitudes on Taiwan in the direction of acceptance of the mainland's position. However, even as economic ties have strengthened, there has been no evidence of any relaxation of the concern over independence tendencies on Taiwan or emphasis on the importance of political negotiations.

The opposite is that unbalanced economic ties increase Taiwan's vulnerability to mainland pressures to abandon its stance. This view has many dimensions. There is the assertion that, as a result of manufacturing on the mainland, Taiwan's industrial infrastructure is being "hollowed out" and will leave the island dependent on China. In addition, given candid statements from Beijing that it wishes to use economics to put political pressure on Taiwan, some argue that a growing Taiwanese business community on the mainland could become a subversive force that would use its wealth and influence in Taiwan politics to force concessions. Finally, there is the view of the economic interests on the mainland being vulnerable to pressure from Beijing through everything from harassment to confiscation or nationalization.

The validity of these assertions is not clear. There is little evidence of the business community on the mainland or its representatives seeking to pressure the government on Taiwan into making political concessions. Moreover, the argument about the "hollowing out" of manufacturing fails to take into account the fact that much of the manufacturing that went to the mainland was no longer profitable on Taiwan and that a significant manufacturing base still exists on the island. Further, at the present stage of development, the economy should be expanding a more internationalized services sector.

The possibility that threats to hold Taiwan's economic interests on the mainland hostage might be used as a way of gaining compliance with Beijing's demands is unlikely due to the repercussions of such an act. In the first place, as we have seen, China remains an export-driven economy despite PRC efforts to turn more purposefully in the direction of domestic-driven demand growth. Taiwanese companies manufacture the bulk of exports in the information and communications field. Interference here would not only be a blow to China's economy but would disrupt a global chain affecting other Asian countries. Moreover, Taiwanese companies are manufacturing for American companies, and the use of economic coercion would affect both US–China relations and global views of China's reliability as an economic partner more generally.

At the local level, Taiwan enterprises and their suppliers are important sources of employment and tax revenue. Anecdotal evidence suggests that, during difficult periods in cross-strait relations, local officials in areas with substantial Taiwanese investment have argued for moderation and sought to reassure business people from the island. National leaders have been similarly restrained and have gone out of their way to reassure Taiwan's investors on the mainland during times of tension.

It thus seems likely that, should Beijing decide to coerce Taiwan to negotiate unification, it would probably not use economic tools. The more likely instrument of coercion would be military power. This possibility is considered in the next chapter.

8 | The Security Dimension ───────

The previous discussion has depicted a cross-strait relationship that embodies a number of contradictory and ambiguous threads. Economic relations have thrived between the two sides for more than two decades, yet efforts to institutionalize that relationship or move it into non-economic areas have encountered difficulties. Beijing has continued to reserve the right to use force should unfavorable trends develop, the most egregious of which would be a declaration of independence by the government in Taiwan – this, despite the fact that both major parties in Taiwan have already declared that the ROC already is a sovereign and independent state (albeit that one party says the state includes the mainland and the other does not). The United States stands by President Clinton's formula that the issues between Taiwan and the mainland "must be resolved peacefully and with the assent of the Taiwan people," but insists that it "does not support" the island's independence – a possible choice by those people.

These contradictions reflect the reality that, while the three major actors appear to accept the current situation in the Taiwan Strait, they all see it as less than ideal – even potentially damaging to their interests. Peace in the Taiwan Strait is a vulnerable structure that persists only because all sides are forced to accept an unacceptable status quo.

But what if this acceptance is challenged? Suppose the leadership in Beijing concludes that any further continuation of the situation of stalemate will result in Taiwan's permanent separation from the mainland and decides to resolve the situation before it comes to that. Or what if some future government in Taipei actually provokes Beijing by pursuing policies suggesting a lurch toward independence? If armed conflict results from these circumstances, what would the response of the United States be?

This is a conflict that no side seeks, but it is one for which all sides are preparing. Indeed, as we shall see, the nature of these preparations adds considerable volatility to any future crisis situation. Not only are there signs of an arms race, but the strategic planning regarding the use of those weapons indicates that any conflict there will not only escalate quickly but also expand beyond the cross-strait area into East Asia.

CHINA: PREPARING ON TWO FRONTS

Preparing for a Taiwan contingency has been a major driver of China's military modernization over the past three decades. Specifically, the events of 1995–6, when the United States first sailed an aircraft carrier undetected through the Taiwan Strait and then sent two carrier battle groups to the area in response to Chinese missile tests, are generally identified as prompting this effort. This modernization intensified during the Chen Shui-bian administration and, despite the improved atmosphere that followed the election of Ma Ying-jeou in 2008, continues until the present time. The result has been a radical transformation in the cross-strait military balance and, with it, increasingly pessimistic appraisals of Taiwan's ability to resist efforts by the mainland to use force to achieve its goals in the strait.

At the turn of the twenty-first century, most commentators considered that Taiwan had a good chance of resisting a military campaign launched by the mainland (Shlapak 2009: 223). Today, that has all changed. According to the 2013 United States Defense Department *Annual Report to Congress*, China has more than three times as many troops in the area as Taiwan (which number 400,000) and possesses not only an imposing submarine force but also various types of surface vessels, which can both assist in amphibious operations and serve as platforms from which to launch missiles against other vessels at sea or in support of land operations (US Department of Defense 2014). Considering only Taiwan's forces, the mainland has also achieved supremacy in the air. It possesses a large number of advanced aircraft that have been purchased from Russia or developed within China, as well as a sophisticated air defense system that extends beyond the Taiwan Strait. Perhaps most dramatically, on the east coast there are approximately 1,600 missiles (both ballistic and land attack), representing the bulk of China's total inventory, which are capable of reaching not only Taiwan but certain American bases in the Pacific. Finally, significant progress has been made by China in areas ranging from cyber and electronic warfare to command and control. In short, without taking account of American capabilities, the military balance in the strait has shifted decidedly in favor of the mainland.

This imposing array of forces performs two functions in the mainland's management of cross-strait relations. The first is as a deterrent – a threat of the consequences that might result from any actions considered unacceptable to Beijing. In this sense, the mainland's military power plays a passive role, standing as a warning of action that might be taken should Taipei act in a provocative fashion. The second function is more active. It is compellence – using the threat of force or its use to compel the government on Taiwan to take certain actions, such as acceding to China's terms for unification, or to desist from unacceptable policies, such as moving toward formal independence. It is this actual use of force and its consequences that is of greatest concern and that has been a significant issue in Sino-American relations for decades.

We have seen that, ever since the ambassadorial talks in the 1950s, the United States has sought Beijing's agreement to abandon the use of force in resolving cross-strait issues. China has refused to do so, arguing, on the one hand, that it has the right to resolve a domestic matter without outside interference and, on the other, that, without the continuing threat of force as the ultimate solution to cross-strait matters, authorities on Taiwan would opt for independence. Over the years Beijing has identified a number of actions short of independence, such as occupation by foreign troops or Taipei's acquisition of nuclear weapons, that might require it to respond with force. However, there have been suggestions that it might resort to coercion under conditions that are even more ambiguous and subjective.

This is the case with both the 2000 White Paper and the Anti-Secession Law of 2005. In the former, Beijing declared that, "if the Taiwan authorities refuse, sine die [indefinitely], to the peaceful settlement of cross-Strait reunification through negotiations, then the Chinese Government will only be forced to adopt all drastic measures possible, including the use of force, to safeguard China's sovereignty and territorial integrity and fulfill the great cause of reunification" (Taiwan Affairs Office 2011). In the latter, it was stipulated that

> In the event that the "Taiwan independence" secessionist forces should act under any name or by any means to cause the fact of Taiwan's secession from China, or that major incidents entailing Taiwan's secession from China should occur, or that possibilities for a peaceful reunification should be completely exhausted, the state shall employ non-peaceful means and other necessary measures to protect China's sovereignty and territorial integrity. (National People's Congress 2005)

A number of ways in which the leadership in Beijing might exercise military pressure have been suggested. A blockade of the island might be declared, disrupting foreign trade, including necessary imports for the island's economy. There could be a demonstration of cyber warfare that would disrupt necessary civilian and military communications. Climbing further up the ladder of coercive tools, Beijing could decide to use its considerable force of short-range ballistic missiles and land-based cruise missiles on vital military and economic targets, which might be followed up by a limited invasion by air or sea. Finally, should all else fail to secure the necessary compliance with mainland demands, the final option would be a full-scale missile attack, possibly accompanied by an invasion.

In the judgment of most military analysts, with the exception of a full-scale amphibious landing, all of these options are presently available to mainland leaders should they decide that deterrence has failed and coercive means must be used to change the behavior of the government in Taipei. However, in contemplating these options, other factors would surely be considered. The first would be the immediate disruption of the complicated strands of production chains of which Taiwan and the mainland are a part (discussed in the last chapter), as well as the probability of curtailment of global economic ties. Another would be the reputational costs for China of resorting to compellence. For, to many observers, how Beijing's leaders handle the Taiwan issue is a litmus test of future international behavior. A resort to force, even if successful, would send an unfavorable signal in this respect. On the other hand, should compellence fail, not only would the reputational costs be high, but the failure to deliver on its decades-long commitment to unify the Chinese nation would undoubtedly impact the legitimacy of the Communist Party and its leadership.

The success or failure of compellence would not be determined simply by the extent of China's military might, as the resistance such efforts would encounter would also be a factor. Such resistance would come in the first instance from the military and the people of Taiwan. The island's military preparedness would determine the extent to which it could defend against or, at least, diminish the full impact of a mainland assault, thus not only minimizing the destruction of the island's defenses and infrastructure but also sustaining the morale of the population to resist. Accomplishing both would also be necessary to allow time for international pressure and/or support from the United States to come to the aid of Taiwan.

Of course, assistance from the United States is not a certainty. The Taiwan Relations Act does not guarantee American intervention, and getting into conflict with China would certainly be controversial within the United States. However, in light of past behavior and statements, a prudent Chinese leadership would have to assume that aid would be forthcoming from the United States, quite possibly including direct military involvement. Should it decide to resort to coercive diplomacy to resolve the cross-strait issue on its own terms, Beijing would have to consider the likelihood of a two-front conflict, as it sought not only to break Taiwan's resistance but to prevent the United States from coming to its aid with supplies or through military intervention.

Chinese military planning has focused on developing what is referred to in US military parlance as an "anti-access/area denial (A2/AD) force" to meet this eventuality (US Department of Defense 2013). This assumes asymmetry between a more powerful American and a weaker Chinese military. Its purpose is to slow the deployment of US forces into a combat area or compel them to operate at a distance from it (A2) and, failing that, to disrupt their operation within the area (AD). The methods employed would encompass a wide range of actions intended to delay or diminish the impact of any intervention to assist Taiwan. According to Chinese military writings, the focus of such an effort would be primarily on deployed American naval and air forces as well as on the communications and command systems that support them.[1]

At sea, the Chinese focus would be on American aircraft carriers. Anti-ship ballistic missiles, sea- and air-launched guided munitions, and submarines would all be mobilized to disable the vessels or keep them at a distance. An advanced and complex anti-aircraft system would be utilized to deal with any aircraft that might be carrier launched as well as aircraft from nearby American bases in Guam, Okinawa, Korea, and possibly Japan. Finally, the expectation would be that cyber warfare would be utilized in an attempt to disrupt intelligence as well as command, control, and communications capabilities. In short, it is assumed that the Chinese would seek to attack on the widest scale feasible in order to diminish the American advantage.

The manner in which this strategy might be implemented is as significant as its scope. Chinese discussions emphasize the importance of preemption, initiative, and surprise as providing additional advantage to the weaker force seeking to keep the stronger at bay. This means that, if a crisis develops in the Taiwan Strait, the Chinese leadership might calculate that it would be best to make a preemptive strike on American forces in the area rather than assume non-intervention. In short, should Beijing take a decision to use coercive means to respond to developments in the Taiwan Strait, the mainland might not wait for evidence of American intervention. It would seek to take advantage of surprise and preemption to compensate for American military superiority.

Put another way, in current Chinese military doctrine, the use of force in the Taiwan Strait is strategically linked to the use of preemptive force against the United States, creating the conditions for a self-fulfilling prophecy of American intervention.

TAIWAN: CREATING A "HARD ROC"

During the period of martial law in Taiwan, the authoritarian KMT shared two beliefs with the communist regime on the mainland: that Taiwan was a part of China and that the cross-strait conflict would eventually be settled by military means. The impossibility of the latter task as well as the efforts of the United States restrained the Taiwan government, although there were certainly moments of crisis. American efforts also persuaded the KMT to highlight the importance of peaceful methods, specifically the attraction of Taiwan as a thriving democracy, to effect a change in mainland policy – the belief that motivated Chiang Ching-kuo toward the end of his life.

Thus, when Lee Teng-hui became president, although nominally holding to the idea of one China, his government committed itself to a peaceful, negotiated approach that was symbolized in May of 1991 by the termination of the Temporary Provisions for the Suppression of the Communist Rebellion – in effect, he declared an end to the civil war. And so, with the mainland continuing to reserve the right to use force, and providing evidence of that determination in 1995–6, the strategic posture of Taiwan shifted to one of defense in the face of China's continuing threat.

This, of course, was not popular with the Taiwan military, for whom the return to the mainland had been seen as a sacred task. However, this gradually changed with the Taiwanization of the military. By the mid-1990s the doctrine of "pure defense" or "all-out defense" had become the strategic orthodoxy. The mission of the military was defined as defense of those areas under the control of the ROC government. Toward the end of the Lee administration, there was a subtle shift to a strategy of "effective deterrence and resolute defense." The emphasis on deterrence suggested a posture that would make the mainland think twice before launching an attack by developing the island's capacity for defensive operations in and across the strait (Swaine and Mulvenon 2001).

This all changed again during the administration of Chen Shui-bian from 2000 to 2008. During the election campaign, all candidates presented strategies that were more offensive in their character, but Chen's seemed the furthest from the earlier, more passive posture. He called for a strategy of "offensive defense" that emphasized "paralyzing the enemy's war fighting capability" and "keeping the war as far away from Taiwan as far as possible." Specifically, this involved a "campaign beyond boundaries" that would "actively build up capability that can strike against the source of the threat." This concept was part of a broader concept dubbed "offensive deterrence" or "preemptive defense." By the time Chen left office, Taiwan's official doctrine had become a more moderate-sounding doctrine of "effective deterrence and resolute defense," which proposed a strong defense that would deter an opponent and punish it should it choose to attack (Swaine and Mulvenon 2001).

As we saw in chapter 5, Beijing looked upon Chen and his party with great suspicion. To what extent Chen and his increasingly provocative policies after 2002 were actually tied to China's military modernization during these years is uncertain. However, it was during the eight years that Chen was in office that the enormous increase in the quality and diversity of China's military grew to the extent that any illusions that Taiwan might prevail in a cross-strait conflict were dispelled. The most dramatic indicator of this growth was the number of land-based missiles pointing toward Taiwan. When Chen took office in 2000, it was estimated that China had 200 short-range ballistic missiles facing Taiwan. As he was about to leave office, he announced that his government believed that the number had grown to more than 1,300.

The shift in the cross-strait military balance was not simply the result of the growth in mainland capabilities. Although one could not expect Taiwan to keep pace with the developments on the mainland, it is probably true that during these years ground was lost due to the security policies of the new DPP administration. As we shall see below, a year after Chen took office, the incoming Bush administration approved the largest arms sales package since the sale of the F-16s in 1992. A wide variety of weapons was offered for sale, and these were intended, for the most part, to respond to the growth in the mainland's military. However, as one author has put it, the Chen administration "dithered," delaying Taiwan's acquisition.[2] Complicating the picture was the fact that the opposition KMT controlled the legislature throughout the eight years of Chen's presidency and there was continual political squabbling over funding the military.

Democracy affected defense spending in another way. Growing popular demands for public services made it impossible for the government to maintain the funding levels that had existed earlier. With the military budget at less than 3 percent of GDP, not only did the cross-strait gap, but the Bush administration, already frustrated with the provocative mainland policies of the Chen government, suggested that the political wrangling over the defense budget raised questions about the administration's determination. Washington hinted that it might be unwilling to help a government that was not ready to help itself. By the end of Chen's presidency, Taiwan had purchased only a portion of what had been offered by the Bush administration, with the politically popular Patriot anti-missile defense system one of the more expensive items funded.

However, in certain areas, the Chen administration made progress. For example, in the annual report for the Ministry of Defense issued during Chen's last year in office, the integration of "industrial, governmental, and academic technological resources" in the creation of indigenous weaponry was highlighted. Taiwan continued to produce an indigenous fighter plane as well as ship-to-ship and surface-to-air missiles. Efforts were also begun to harden communications centers and air force facilities to ensure their survival during any missile attacks, and, most dramatically, the construction of a hangar in the mountains in the center of the island was completed. Moreover, despite the sharp exchanges between Chen and the Bush administration, defense cooperation between the United States and the ROC was considerably expanded during these years.

When the Ma administration took office in 2008, it was obvious that the cross-strait balance had tipped decisively in favor of the mainland. Ironically, Taiwan was faced with a problem similar to that confronting the mainland: devising an asymmetric war-fighting strategy for dealing with a militarily superior opponent. An alternative that was seriously considered by Ma's advisers was one proposed by William Murray at the Naval War College in the United States (Murray 2008).

Murray, a retired naval officer, argued that, given the mainland build-up, Taiwan could no longer hope to achieve parity with China's military. Rather, it needed to accept the role of a defender and adopt what he called a "porcupine" strategy to allow the island to sustain an extended defense "for weeks or even months." This would entail hardening and dispersal of command centers, as well as weaponry and the storage of supplies, and the strengthening of utilities so as to survive a blockade. Most importantly, Murray argued that such a strategy required that weapons high on Taiwan's wish list, such as advanced fighter aircraft and submarines, should be de-emphasized in favor of such defensive weapons as anti-ship missiles, helicopters, and mines that could not be easily destroyed or, in the case of planes, be stranded on the ground or in the air as a result of runway destruction by Chinese missiles. Such a defense, Murray argued, would make the mainland think twice about attacking and, in the event of a strike, demonstrate to Washington the island's determination to resist and provide a breathing space for the United States to decide whether to come to Taiwan's aid.

In 2009, the Ministry of Defense produced its first *Quadrennial Defense Review*, which served as an introduction to the Ma administration's strategic doctrine, designated as "Hard ROC." This term signified "a rock-solid and impregnable defensive force that, by implication, could not be dislodged or breached by a numerically superior enemy force during an attempt to attack or invade ROC territory." Reversing the order of the slogan used during the previous administration, this was characterized in the report as a strategy of "resolute defense and credible deterrence" (Ministry of National Defense 2009). The logic was spelled out in the 2013 quadrennial report, which asserted that "resolute defense" would involve a "fortified" defensive capacity intended to "sustain [the] enemy's first strike, avert decapitation, and maneuver forces to carry out counterattacks and sustainment," while training and combat preparation would force "the enemy to consider the costs and risks of war." Perhaps the most striking element in the new doctrine was the commitment to a smaller, all-volunteer military force. However, as of this writing, this had still not been successfully achieved (Ministry of National Defense 2013).

In other respects, it seemed clear that, as this strategy evolved during the years between the two quadrennial reports, the doctrine represented neither a return to the purely defensive doctrine of the 1990s nor a total acceptance of Murray's porcupine strategy.[3] To be sure, much of the thinking of the latter could be found in the greater emphasis on hardening, dispersal, the development of mobile missile launchers, smaller naval attack vessels, and a variety of anti-ship missiles that would be more likely to survive an enemy's first strike and provide for the island's defense. However, there have been other trends that suggest a doctrine that goes beyond that of mere passive defense to offshore defense and even attacks on the mainland.

This has been particularly obvious in the development of indigenous weapons that have offensive capabilities. For example, during Ma's second term, Taiwan unveiled the "Wan Chien" (ten thousand swords) missile or "smart bomb," which carried 100 warheads and could be launched from the air to attack harbors, airfields, anti-aircraft positions, etc., on the mainland. In addition, Taiwan developed its first land-attack cruise missile capable of reaching the mainland. It is transported on a mobile launcher hidden inside a vehicle disguised as a delivery truck. Finally, there were reports that Taiwan was deploying a "Yun Feng" (cloud peak) missile with a range of more than 1,000 kilometers. The trend in the Ma administration has clearly been to expand production of defense tools at home (Defense News 2014).

One reason is economic. Foreign purchases are both expensive and politically difficult in a new democracy, where people demand a wide range of public goods. However, it is also because of Taiwan's limited access to the international market in arms. Strong mainland protests as well as retaliatory actions have discouraged foreign sales. This has resulted in greater domestic development and production as well as increased reliance on American weaponry, which is not only expensive but subject to Washington's concerns regarding the development of offensive weapons.

Still, the relationship with the United States remains the most important factor in Taiwan's defense. As was the case in the 1950s and 1960s, American-supplied arms have been an integral part of Taiwan's military. In the period since normalization, American administrations have sought to strengthen Taiwan's defensive capability in a manner that minimizes the anticipated and inevitable impact on relations with Beijing. Moreover, as was the case during the Chen Shui-bian administration, Washington has made it clear that it considers arms purchases as a barometer of the island's commitment to its own defense, which, in turn, influences the nature of the American commitment.

From 2000 to 2011, the United States delivered approximately $8 billion worth of arms and completed agreements valued at about $9 billion.[4] The largest items were the PAC-3 anti-missile system, naval destroyers, attack helicopters, early warning radar, and anti-submarine aircraft. However, there were also examples of limits on sales. The most prominent of these have been a continuing saga concerning submarines and the refusal by the administration of George W. Bush to sell Aegis-equipped Arleigh Burke-class destroyers and by that of Obama to sell newer models of the F-16 fighter plane (the F-16 C/D).

In 2001 the Bush administration agreed to sell eight submarines to Taiwan to replace the island's fleet of two Dutch submarines purchased in the 1980s and two World War II American submarines used for training. The problem was that the United States did not produce such submarines, and there was resistance within the US navy to constructing anything but nuclear submarines. There was some talk of buying foreign submarines or technology and reselling them to Taiwan, but potential sellers were concerned with the backlash from the Chinese. Negotiations have dragged on since with little progress, as the project has become entangled in budgetary squabbling in Taiwan and whether such submarines might be produced on the island, and at least one Obama official indicating that they might be considered offensive in nature.

The security relationship with the United States extends beyond arms sales into other aspects of defense cooperation. Yearly talks with Taiwan on military and strategic matters (the so-called Monterey talks) have taken place since 1997. Taiwan officers attend educational institutions such as the Naval War College, receive flight training at American airbases, and have access to the meetings at the Pentagon as well as the headquarters of the Pacific Command. Officials in the Ministry of Defense (including the minister and the chief of the general staff) have traveled to the United States for meetings with American officials and representatives of American defense industries. There are similar visits by Taiwan intelligence officials.

Since 2002, the American Institute in Taiwan has had an active duty military officer at the rank of colonel who is charged with supervision of the assistance programs. In addition, there have been delegations of serving and retired American personnel at military exercises on Taiwan, while the Defense Department has conducted numerous assessments of various aspects of the island's military preparedness. Finally, of course, there is a considerable exchange of information and advice. Much of the recent reform in the Taiwan military, ranging from dispersal, to hardening, to a reduction of the usual preference given the army, has reflected the American influence.

There is, of course, another potential role for the United States: intervention to assist Taiwan in the face of a mainland attack. In 2010, at a time when his administration was successfully negotiating agreements with the mainland, Ma spoke of the chance of conflict between Beijing and Taiwan as being the lowest in sixty years. Contrasting his stewardship with that of Chen Shui-bian, he went on to say: "We will continue to reduce the risks so that we will purchase weapons from the United States, but we will never ask the Americans to fight for Taiwan…this is something that is very, very clear" (Evans 2010).

Although Taipei defense writings contend that the purpose of the "Hard ROC" defense is to create a Taiwan so strong that the costs of invasion would be prohibitive and thus act as a deterrent, in the spring of 2014 a Ministry of Defense report acknowledged that China would "be able to take Taiwan by force before the end of 2020," and numerous exercises have been held to determine how long the island could hold out if attacked. The implicit assumption is that a sturdy defense would not only demonstrate the island's determination to resist but would give the United States time to come to its assistance.

As we have seen, Beijing also appears to be planning on the likelihood of conflict with the United States as a consequence of the use of force in the Taiwan Strait. Moreover, plans for countering that intervention appear to increase the likelihood that any such conflict will escalate to become a Sino-American conflict.

THE UNITED STATES: DUAL DETERRENCE

Since June of 1950, when President Truman committed the Seventh Fleet to the Taiwan Strait, the United States has been committed in some manner to the defense of Taiwan. From 1955 until 1980, it was formalized in a mutual defense treaty. Since then this contingency has been based on the Taiwan Relations Act. However, in all of these cases American support was neither assured nor automatic.

As we have seen in the earlier chapters, several American administrations have viewed this relationship as a burden and even a danger. Thus during the 1950s and 1960s the concern was that the defense treaty would entrap the United States in a war with the mainland. Indeed, President Truman's dispatch of the Seventh Fleet was designed to prevent provocation by either side. Later the Eisenhower administration used the treaty and the connections with the Taiwan military as much as a means to restrain Taiwan as to defend it. Eventually the growing power of China, and Taiwan's abandonment of the option of using force to regain the mainland, shifted the focus of the defense relationship to the importance of deterring mainland aggression against Taiwan.

That became especially clear as a result of Taiwan's democratization in the 1990s. The legitimation of public discussions of independence, and the growing realization on Beijing's part that the China policy under Lee Teng-hui lacked the pro-unification orientation of the old KMT, stoked cross-strait tensions. The 1995–6 confrontation resulted in what appeared to be an unambiguous display of American words and actions intended to deter the mainland from attempting to use force to change Taipei's behavior.

Yet Lee's behavior was also a matter of concern to the Clinton administration. It was during this crisis that his administration announced the defense policy of "resolute defense, effective deterrence" that implied possible attacks on the mainland. This raised some concerns in Washington that its support might encourage provocative behavior by Lee that could escalate the already tense situation, and President Clinton reiterated that the United States did not support independence. These concerns became more explicit during the cross-strait tensions that resulted from Lee's "two states theory" of 1999, when a highly placed State Department official reminded Taipei that the United States "would not support unilateral changes in the overall status of PRC–Taiwan relations." In short, by the end of the Clinton administration, American security policy in the area was returning to an approximation of the policy of the Eisenhower administration, which had been one of deterring a Chinese attack on the island while seeking to prevent provocative actions by the ROC that might entrap the United States (Tucker 2009: 222; Suettinger 2003: 382).

Some have labeled the policy developed in response to this situation "strategic ambiguity." It assumes that leaving leaders on both sides of the strait uncertain as to what the American response in a particular conflict might be avoids the danger that too strong a commitment to the defense of the island might prompt Taipei to do something that would cross a mainland red line, and that too weak a commitment might encourage the mainland to resolve the conflict by force. It is best, the argument goes, to leave all sides uncertain regarding the American response. As then Assistant Secretary of Defense Joseph Nye told the Chinese when asked what American policy might be should a conflict break out, "We don't know, and you don't know" (Suettinger 2003: 244).

However, Richard Bush, who was chairman of the board of the American Institute in Taiwan, has suggested that this is too passive a characterization of American policy as it developed after the 1990s. He proposes the term "dual deterrence," meaning that the United States combines threats regarding the consequences of certain actions with assurances of support for each side's position should they abstain from provoking conflict in the strait. Thus, Taiwan benefits from its relationship with Washington in exchange for not threatening the status quo in the area, while China is assured that the United States will not support Taiwan's independence in exchange for the mainland refraining from the use of force. "In effect," he argues, "Washington's message to both Beijing and Taipei is that it will defend Taipei under some circumstances and not others" (Bush 2005: 258–65).

It was during the administration of George W. Bush that this message of conditionality was clearly articulated in reaction to the policies of Chen Shui-bian (Kan 2014: 29). One aspect of the policy was a response to the political bickering in Taiwan over the budgetary allocations necessary to purchase American arms. The message was very simple: Taiwan would be expected to carry its share of the burden of its defense if it expected American support. The Chen administration was thus put on notice that the extent of American support could not be assumed. It would depend on reciprocal efforts by Taiwan.

There was a second element of conditionality introduced during the Bush administration. In 2005, the president, who said earlier that the United States would do "whatever it took to help Taiwan defend itself," warned that, "if China were to invade unilaterally, we would rise up in the spirit of the Taiwan Relations Act. If Taiwan were to declare independence unilaterally, it would be a unilateral decision that would change the U.S. equation" (Kan 2014: 24). Throughout the presidency of Chen Shui-bian, administration officials were careful not to create the impression that the American commitment to Taiwan's defense was unqualified. Such a posture, it was hoped, would deter Chen from sparking a conflict with the mainland by his policies. It was dual deterrence in action.

As we saw, when Ma Ying-jeou became president in 2008 he was determined to address the problems of Taiwan's relations with the United States. His pledge not to be a "troublemaker" was an attempt to ease the concerns of both the mainland and the United States. He made it clear almost immediately upon taking office that he would seek to increase military spending to 3 percent of GDP and would intensify efforts to purchase American arms. Moreover, the "Hard ROC" defense strategy clearly reflected the efforts of the Pentagon to encourage Taiwan to adjust its defense policies in the direction of an asymmetric approach.

On the other hand, Ma's efforts to undo the results of Chen's provocative mainland policy did cause some concern in Washington. As we saw in the last chapter, it prompted a wave of speculation regarding the impact that the improvement in cross-strait relations might have on Taiwan's relationship with the United States. The closer relations with the mainland were seen by some as a sign of a further movement toward the mainland that would ultimately benefit Sino-American relations by removing a major encumbrance to better relations. Others were alarmed that the drift toward the mainland would damage American interests in the area, and, among these, some argued for enhanced support for Taiwan so that it could resist Beijing.

This uncertainty regarding the direction cross-strait relations were taking was one factor that renewed congressional activism on policy toward Taiwan. By President Obama's second term, there were members of Congress who, as had been the case during the Clinton administration, attempted to use the TRA as a tool to change what was seen as an insufficiently supportive position of the island. Besides the occasional motion introduced to commemorate an anniversary or upgrade the status/image of Taiwan–US relations, their major effort, as was the case earlier, has been to advocate for Taiwan's perceived interests in the arms sale process. More broadly, Taiwan's supporters have sought to introduce legislation to direct the sale of specific weapons systems. In 2013, they introduced a wide-ranging bill, the Taiwan Policy Act of 2013, which, like the earlier Taiwan Security Enhancement Act, sought to upgrade the nature of the relationship on a wide front, ranging from support in international organizations to arms sales. The attempt died in committee in the Senate (House Committee on Foreign Affairs 2013).

The Taiwan–American security relationship has returned to a more even keel since Ma took office. The issue of Taipei's arms budget continues to be a sensitive issue, as the goal of 3 percent of GDP devoted to defense has not been achieved during his presidency. However, after some delay, in 2010 and 2011, the administration of President Obama notified Congress of proposed arms sales agreements amounting to $6.4 billion and $5.9 billion had been reached with Taiwan. The mainland has continued to object that these sales interfere in China's internal affairs and diminish Taiwan's incentive to negotiate. However, the position of the Obama administration has been that, to the contrary, the security relationship, and particularly the arms sales, complement, rather than contradict, stability in cross-strait relations. Like the Bush administration, it contends that these sales provide Taiwan with the strength and the self-confidence needed to pursue détente with the mainland.

Where current policy has differed from that of the Bush administration is that deterring the island from provoking the mainland has disappeared from the security discourse between the United States and Taiwan. Thus, while deterrence of Taiwan from provoking conflict undoubtedly remains American policy in the cross-strait area, since 2008 the primary focus has clearly been on the proper response to China's growing military power, as well as on the shift that has taken place in the military balance – not only in the strait area but also in the Western Pacific as a whole.

Of course, this broader focus is nothing new. Since the end of World War II, American policymakers have always considered the relationship with Taiwan within the context of the nation's broader strategic posture in that area. Moreover, the TRA had explicitly placed the relationship with Taiwan in the context of a stated American interest in peace and stability in the Western Pacific as well as establishing the requirement that the United States maintain a military presence sufficient to protect that interest. Thus, it was inevitable that when, in 2011–12, the United States announced a "rebalance" toward Asia, the question of its relationship to the Taiwan Strait would be raised. The rebalancing was not only about military matters, and the concern that motivated it was not only the possibility of a confrontation in the Taiwan Strait (Sutter et al. 2013). However, discussions of the military component have focused on the possibility of a Sino-American confrontation in the Taiwan Strait, and the conclusions are ominous.

As we have seen earlier in this chapter, the dramatic growth of the Chinese military has been accompanied by two fundamental principles: that the prevention of Taiwan independence is a preeminent task for the PRC military and that planning for that task must assume the likelihood of American intervention in support of the island.

Moreover, it is recognized that, while military superiority can be achieved vis-à-vis Taiwan, China is unlikely to achieve a similar status in relation to the United States. For this reason, the mainland is developing an asymmetric strategy for dealing with the possibility of intervention based on a strategy of first deterring and then defeating American attempts to bring the full force of its military power to the conflict – the area denial strategy.

As early as 2009, planning was authorized for a response to the problem of area denial under new circumstances. The "problem" to be addressed by what became known as "Air-Sea Battle" was that "adversary capabilities to deny access and areas to U.S. forces are becoming increasingly advanced and adaptive. These A2/AD capabilities challenge U.S. freedom of action by causing U.S. forces to operate with higher levels of risk and at greater distance from areas of interest." Specifically, Pentagon planners assumed that, in implementing an area denial strategy, the adversary would strike with little or no warning, attacking both forces in the area and those in the territories of the United States and its allies that were involved in operations. It assumed that "all domains will be contested – space, cyberspace, air, maritime, and land" – and, finally, that none of these domains can be "completely ceded." This description is, of course, consistent with the Chinese posture described earlier in the chapter.

The response to these challenges was said to be to develop integrated, joint forces that, with short warning time, could collaborate to carry out "in-depth" attacks on the enemy. The goals of these attacks would be to disrupt command, control, surveillance, intelligence, etc., facilities that could be used to identify American forces, "destroy A2/AD platforms and weapons systems," and "defeat" weapons that have already been deployed. This sequence was referred to as the enemy's "kill chain," which goes from finding a potential target to directing a weapon and launching it, and the essence of Air–Sea Battle is to disable the chain (US Department of Defense 2013).

Despite occasional denials that this strategy is directed at China, the nature of the approach, the comments of defense specialists, and, most important of all, the assumption of Chinese observers is that this was, in fact, the strategy of the United States to cope with the Chinese military build-up and adoption of an asymmetric strategy. Moreover, in congressional testimony and the writings of defense specialists, the challenges of implementing Air–Sea Battle against a Chinese adversary were elaborately developed. In these discussions, there is considerable emphasis on the need to develop counter-measures to enemy attempts to disrupt command and control through cyber warfare while defenses against submarines, anti-ship missiles, and anti-aircraft are being developed. Finally Air-Sea Battle clearly assumed that attempts to implement an anti-access strategy would involve attacks on American bases in the area. In the case of Taiwan, this would mean not only Guam and Okinawa, but possibly also Japanese territory (Cliff et al. 2007; ESRC 2014; O'Rourke 2014).

The logic of Air-Sea Battle thus demanded that, in order to assure the integrity of United States facilities and forces, there would have to be attacks on the first point in the "kill-chain" – facilities on the Chinese mainland. While it has been the policy of the United States to discourage Taiwan from developing offensive weapons to strike the mainland, ironically, the logic of Washington's own response to attempts by the Chinese to disrupt American intervention implied such strikes. More than that, Air–Sea Battle placed a premium on early strikes on mainland facilities to limit the danger that anti-access strategies might succeed.

Although, in early 2015, the United States Defense Department ceased using the term "Air-Sea Battle" and reorganized its office, current planning intended to deal with anti-access threats appears to remain largely the same. Thus here is the source of the danger that lurks in any crisis in the Taiwan Strait area. Both the United States and China are developing strategic doctrines that emphasize early action and attacks that extend into bases and other supportive installations on the territory controlled by the other side. Such a situation creates, in the first place, a temptation to preempt the adversary and, by doing so, gain the initiative, an important element in Chinese military doctrine but also implicit in the American response. This could produce a self-fulfilling prophecy on the mainland side whereby the Chinese assumption that the United States will come to the aid of Taiwan encourages early strikes on American forces – actions that would only assure that the United States would have no other choice but to intervene.

Related to this is a second implication of the juxtaposition of these two strategic doctrines: the difficulties of limiting the scope and intensity of a conflict in the cross-strait area. Should the Chinese decide to "punish" a future Taiwan government for some unacceptable action by a show of force, or should there be an incident that sparks a low-level confrontation between the mainland and either Taiwan or the United States, it would be quite difficult to restrain the potential danger of escalation.

CONCLUSION: THE DANGERS OF DEADLOCK

The Strait is a dangerous place where the fundamental interests of the United States, China, and Taiwan are in sharp conflict. It is clear that none of these actors wants the differences that divide them to spark an armed confrontation. Yet, as we have seen in this chapter, all are preparing for the worst.

For China, an armed conflict in the Taiwan Strait would not only likely lead to a costly conflict with the United States but, win or lose, would damage the nation's international reputation, which has been so assiduously cultivated as part of China's "peaceful rise." For Taiwan, conflict is nothing less than an existential threat. The island and the waters around it would undoubtedly be a principal battlefield. With a defense strategy that is based on the assumption that the adversary will attack first and that the punishing strikes will continue until help comes (if it does), there is little incentive to provoke confrontation. Finally, for the United States, the growth of the Chinese military in the last decade has considerably changed the calculus regarding not only the possibility of a Sino-American conflict that might develop in the area but also the nature of that conflict. As American analysts openly admit, the asymmetric, anti-access strategy and the Chinese capabilities standing behind it will make any confrontation a dangerous and costly affair that could easily spread beyond the Taiwan Strait and bring about the first large-scale conventional war between two nuclear powers in world history.

And so it is obvious that all three actors in the strategic triangle of the Taiwan Strait have reasons to avoid conflict. However, whether because of concern for the party's legitimacy in China, or the strong distrust of the mainland in Taiwan, or the concern of the United States to ensure the maintenance of peace and stability in Asia, none of the actors seems ready to make the concessions necessary to lift the threat of armed conflict that hangs over the strait. In the concluding chapter we elaborate on this shaky status quo.

Conclusion

This analysis began with the assertion that the parameters and content of the policies pursued by the major actors in the Taiwan Strait have been the product of history. During the period from the end of World War II until the 1990s, cross-strait relations were largely an issue in Sino-American relations. It was during this time that the fundamental positions of each side which remain largely in force today were defined. In these years, the irreconcilable nature of the American and Chinese positions was demonstrated by conflict, misunderstandings, and shaky compromises.

By the last decade of the twentieth century, the focus shifted to relations across the strait between Taiwan and the mainland. The democratization of Taiwan ended the island's wartime stance and fostered a broad range of exchanges across the strait. However, democratization meant popular reaction against previous policies of the KMT. This was directed at weakening the assumption of Taiwan's political connection to the mainland and seeking to establish the island's distinctive, sovereign status.

Since democratic Taiwan began to reshape relations, the United States and China have found themselves interacting with each other and the island's government largely on the basis of the principles established earlier. We have seen that the challenges to Sino-American relations were the greatest during the administration of Chen Shui-bian. His efforts to establish greater political distance from the mainland and to enhance Taiwan's international sovereign status were seen by the mainland as upsetting the status quo as they defined it – that Taiwan was a part of China. For the United States, the danger was a different one. When President George W. Bush criticized Chen for upsetting the status quo, he had in mind a situation of peace and stability in the Taiwan Strait and a concern that military action by the mainland would entrap the United States. Chen's policies, thus, produced the unusual result of some parallelism in the strait policy of the two powers.

When Ma Ying-jeou became president, relations took a very different turn. In contrast to Chen, he took a position that, ironically, seemed consistent with the definition of the status quo held by China and the United States while also being consistent with public opinion in Taiwan. For Washington, he promised not to be a "troublemaker"; for the mainland, he accepted the concept of one China while differing on its definition; and, for the Taiwanese public, he asserted that the Republic of China was already an independent, sovereign country.

As we have seen, the change in cross-strait relations was dramatic and new. During the early years of the Ma administration commentators predicted various outcomes for the new stability in the Taiwan Strait. Some saw Taiwan drifting inexorably into the arms of the mainland but differed over whether this would be a blow to American interests in Asia or the removal of a major irritant in Sino-American relations. Others, more cautiously, saw the beginnings of a process of institution-building that might lead to a more constructive and peaceful future emerging out of the new status quo.

As the transition from the Ma Ying-jeou administration begins, it appears that these outcomes have not materialized. Rather a very different kind of status quo has emerged.

THE AMBIGUITY OF "STATUS QUO"

As we have seen, all three actors in the Taiwan Strait define the status quo in a manner that supports their positions. This renders the term unhelpful as a descriptor of the current state of cross-strait relations. However, the term can be useful if we explore the implications of the concept rather than its content.

The term status quo simply refers to things as they are. It tells us nothing about stability or durability.[1] The status quo is robust or peaceful when the parties agree on the rules of the game, but it is unstable when one (or more) of the parties involved is forced to accept suboptimal situations because favored options are not possible. Finally, it is conflicted if one (or more) of the parties considers the status quo to be unacceptable and pursues disruptive policies. In a condition of a robust status quo, all parties are satisfied with conditions and are committed to their indefinite continuation because it is consistent (or at least not inconsistent) with their objectives, while the third condition is exactly the opposite, as competing and irreconcilable objectives generate conflicting policies. The middle alternative is an approximation of Leon Trotsky's "neither war nor peace." It is a condition of stalemate or deadlock where one (or more) of the parties accepts a status quo that doesn't conform to their ultimate goals but is not prepared or able to challenge or disrupt it.

Why would one or more of the actors continue to accept a status quo that falls short of their ultimate objective? One reason might be that the less than optimal situation would be seen as possibly leading to one that conforms directly with ultimate goals. A second rationale might be that external constraints make the cost of opposing such a condition prohibitive. The final reason for accepting a suboptimal situation would be for short-term gains or needed stability despite concerns that the continued maintenance of such a status quo might damage or erode the chances of arriving at a favorable, or more permanent, solution at some later point. In such a situation, the equilibrium becomes fragile and the status quo might transition to conflict.

At present, the situation in the Taiwan Strait is as stable as it has been at any time in the post-World War II period. However, given the past history of the area and the objectives of the actors involved, it cannot be considered a robust status quo. It is rather closer to the middle alternative. The most important question to be addressed is whether it is moving in the direction of a more stable status or toward a more conflicted one; and this, in turn, depends on the level of confidence that the actors have regarding whether the current policies will advance or damage the achievement of their ultimate objectives.

COMPETING OBJECTIVES

As we have seen, the fundamental goal of the Taiwan policy of the People's Republic of China is to incorporate Taiwan into "China" – a goal informed by both a sense of historical mission and strategic calculation. Since the 1970s, the emphasis has been on the preference for peaceful means to accomplish this end. This has been the rationale for mainland efforts to expand economic and cultural links in order to win the "hearts and minds" of the Taiwan people and, by so doing, create a web of relationships intended to enhance the sense of common fate and strengthen interdependence. Moreover, China's leaders have insisted that, ultimately, any acceptable settlement of the island's status must be based on the "one China principle," defined as "there is only one China in the world; the mainland and China both belong to one China; and China's sovereignty and territorial integrity are indivisible." They seek to diminish the salience of a distinctive Taiwanese identity and the continued existence of independence sentiment on the island by offering the island's people participation in Xi Jinping's future "China Dream."

Finally, the mainland seeks to limit any international activity that might suggest the island's separate sovereignty. It is particularly anxious to end American support for, and sale of arms to, Taiwan. As has been the case since the 1950s, this is considered to be interference in China's domestic affairs and encouragement to the government on Taiwan to resist mainland blandishments. As a result, the PRC has dramatically enhanced its military capabilities, not only to deter moves toward independence by Taiwan but also to frustrate American efforts to come to Taiwan's aid should a conflict develop.

On Taiwan, despite deep political differences, the two major parties agree that the present constitution of the Republic of China is the basis for the island's status as a "sovereign, independent state," although, as noted earlier, they differ sharply over whether ROC sovereignty extends to the mainland. This view is responsive to the prevailing mood in public opinion surveys, which reflects the desire to maintain a political system and international presence distinct from that of the mainland for at least the time being. Moreover, although relations have not always been smooth, the ties with the United States remain Taiwan's most important link to the international system.

Against this background, Taiwan administrations since the 1990s have been under pressure from the public to enhance the nation's international profile and to resist acceptance of Beijing's "one China principle." However, the pressures of the business community have been equally strong in advocating the enhancement of economic ties with the mainland – a policy which has been difficult for any government to control, but which has also served to keep the dialogue with the mainland going and to increase the costs to Beijing of any conflict in the Strait.

Turning to the United States, we have seen that the primary objective of administrations in the post-war period has been to avoid entrapment in any cross-strait conflict – especially due to entrapment by Taiwan's actions. American policy explicitly supports the present calm in the area by encouraging cross-strait dialogue toward the ends of peace and stability. Washington complements this support with an agnostic position regarding the outcome of differences, instead emphasizing the process – peaceful negotiation and mutual agreement – while maintaining that any outcome would be acceptable. It has assiduously avoided any suggestion that it would participate in negotiations either as a mediator or as a guarantor of any agreement.

Yet this non-involvement on the part of the United States does not signal indifference to developments in the relations between the mainland and Taiwan. In the first place, since the Truman administration, Washington has been torn between the need to deal with the reality of the PRC and the pressures (whether domestic or international) not to appear to be abandoning Taiwan. To this end, it has sought to maintain a "balanced" policy that leans neither so far to the side of Taiwan that it angers the mainland, or gives Taipei the confidence to provoke the mainland, nor so far to the side of the mainland that it gives the impression of acquiescing to Beijing's pressure on Taipei or appears to abandon Taiwan. And the result of this avoidance of seeming to take sides has resulted in the patchwork of commitments and agreements with both sides that are incomplete, ambiguous, or even contradictory and, most of all, insufficient to win the trust of either side.

Secondly, the policy of dual deterrence that is intended to maintain the peace and stability in the strait confronts the United States with the Hobson's choice of loss of credibility, should it decide not to act, or direct involvement in the controversy, should it decide to act in accordance with the doctrine.

ASSESSING STABILITY IN THE TAIWAN STRAIT

The goals of the actors in the Taiwan Strait are clearly asymmetrical – unification on the mainland side and sovereignty for Taiwan – and there has been little inclination on either side to compromise these objectives in the interest of resolving the conflict. This would suggest that the prevailing status quo is not likely to move in a "robust" direction. However, under what conditions might there be movement in the opposite direction, toward a conflicted one? Paradoxically, the very same policies that are sustaining the present shaky equilibrium could contribute to such a negative trend. Specifically, it is possible for both sides to come to believe that the unintended consequences of the policies that are sustaining the present equilibrium may, in the long run, have the effect of damaging their ultimate objectives. In this case, one side or the other might risk challenging the existing status quo rather than sacrifice those objectives for the sake of temporary equilibrium.

To understand this paradox, one has to examine the development of the status quo in the Taiwan Strait during the administration of Ma Ying-jeou. As we have seen, after 2008, despite remarkable progress in building institutions to manage certain aspects of cross-strait relations, domestic politics in Taiwan considerably slowed the process and blunted Ma's efforts to move the agenda ahead. Polls on the island reflected growing concerns regarding the economic and political impact of agreements with the mainland as well as growing identification with Taiwan and commitment to maintaining the present status quo of separation. The student occupation of the legislature in the spring of 2014 (the Sunflower Movement) demonstrated the growing misgivings in Taiwan regarding the deepening of relations. Finally, while they did not represent a referendum on cross-strait policies, the DPP victories in local elections at the end of 2014 certainly demonstrated the weakness of the KMT. Although the DPP understands the importance of formulating a positive and realistic approach to mainland policy to satisfy the United States and the voters of Taiwan, it is not likely that the party will come up with a formula comparable to the "1992 Consensus" that would recognize Beijing's demand that it accepts the concept of one China.

The mainland has maintained a steady course in the face of these adverse developments on Taiwan that has made a major contribution to maintaining the status quo. Although it continues to take an uncompromising stance toward the DPP and, at times, publicly reminds Taiwan that its determination to achieve unification has not been diminished, there have also been statements of a willingness to be patient, despite the signs of growing resistance on Taiwan. Indeed, despite evidence of growing disillusionment with the effectiveness of mainland blandishments, the Chinese leadership has concluded that there is more work to be done in winning the "hearts and minds" of the Taiwanese people. For example, after the 2014 elections, there was a greater emphasis on the need to work in the future with youth and small-business people on the island. Moreover, despite the little progress that had been made in moving into new areas in cross-strait relations, the mainland continued to explore possibilities for negotiation in less controversial areas, apparently largely for the sake of maintaining momentum. The leadership in Beijing is apparently still confident that, given its military deterrent and economic influence, time is on its side.

Yet, even amid this evidence of patience and confidence, there have been indications of concern that the longer relations continue to move at the present glacial pace, the greater the danger that other trends may develop that will frustrate the mainland's goal of unification. There have been critical statements complaining that Taiwan is getting the bulk of the advantages from trade agreements; that de-Sinification and growing Taiwan-centric consciousness will increase as the separation from the mainland continues; that the direction of the present trends in Taiwanese public opinion suggest not only that the DPP might once again be in power, but that even a victorious KMT would meet constraints in implementing mainland policy; and, finally, that the result of accepting the present slow progress in cross-strait relations might result in Taiwan achieving what has been referred to on the mainland as "peaceful separation" – *de facto* independence. In this view, accepting the present status quo not only delays the realization of Beijing's ultimate goal but might also be fostering conditions that will, in the end, frustrate this goal or make its achievement more difficult or costly. There is no evidence that this view dominates, but it continues to animate the approach of some on the mainland.

On the Taiwan side, as noted, there is similar ambivalence. The administration of Chen Shui-bian had demonstrated how little room for maneuver Taiwan had in seeking to enhance its sovereign status. The new tone in cross-strait relations and the support received from the United States when Ma Ying-jeou was elected were welcomed. Indeed, the concern that this improved relationship with the mainland would not be sustained by the DPP was an important cause of its defeat in the 2012 election. Yet, at the same time, when Ma Ying-jeou raised the question of a peace treaty with the mainland, the sharp public reaction caused him to beat a hasty retreat, and growing public concern regarding the dangers of expanding ties with the mainland amid calm in cross-strait relations became a major theme in his second administration.

Whether articulated as anxieties regarding growing international isolation, economic dependence, or the creation of a pro-mainland constituency, these dangers are clearly present in public opinion polls and, of greater political significance, at the forefront of the concerns articulated by the DPP. In the minds of many, even the limited economic agreements that have been negotiated so far provide the mainland with opportunities to expand its influence through increased investment opportunities, policies favorable to selected constituencies (such as farmers), and interference in the electoral process. From this perspective, any broadening of the scope, or even the continuation, of the current stable status quo carries with it threats to the long-term goal of maintaining Taiwan's identity and sovereign status.

In short, there is a need to be cautious in assessing the durability of the present status quo in the Taiwan Strait or the course that events might take in the future. The present stability is tenuous. Both sides still hold to their basic irreconcilable positions and are aware that continued stability might actually weaken their position. However, thus far they have been willing to discount the unfavorable impact of current policies in the interest of maintaining the current status quo.

The ominous question hanging over the strait is whether they will be willing to continue to do so in the future.

One obvious factor favoring acceptance of the present status quo is the cost to either side in seeking its ultimate objectives. For China, there would be considerable damage to its economy and international reputation in using force to resolve the conflict. For Taiwan, the Chinese deterrent and the generally tepid international support that could be expected from any move to change the status quo serve as constraints. However, in both cases, the dual deterrence policy of the United States is considered to be the most important factor, constraining each side from upsetting the fragile equilibrium provided by what they consider to be a less than satisfactory situation. And so we return to the triangular nature of cross-strait relations and the involvement of the United States in a controversy with its origins in the Chinese civil war more than six decades ago.

We have seen that, at several junctures, it appeared as if the United States would be able to end this involvement. For example, in the months before the outbreak of the Korean War, President Truman seemed ready to do so. Later, during the period from the Nixon–Kissinger demarche until formal recognition, it was assumed that the virtual abandonment by the United States would create pressure on Taiwan to settle with the mainland (Kissinger's "handwriting on the wall").

Yet, in all these instances, efforts to end or minimize involvement have failed. In the case of the Korean War, one might consider the failure of the United States to withdraw from the controversy to be the result of an historical accident. However, since the diplomacy of John Foster Dulles, the United States has sought to fashion a successful policy toward the mainland while retaining ties with Taiwan. It has been the need to balance these two irreconcilable goals that has locked the United States into involvement in cross-strait relations. However, of equal significance are the terms of that involvement. The effort to preserve or enhance Sino-American relations, while not abandoning Taiwan, has created policy commitments that involve misleading or ambiguous statements, unilateral interpretations of documents, and seemingly contradictory policies. These commitments are concretely manifested in the parallel relations maintained with both Taiwan and the mainland. They are expressed in the policy characterized as "dual deterrence," which, in essence, upholds the present status quo by refusing to support Taiwanese independence and cautioning the mainland against using force to secure its goals.

This puts the United States in a difficult position. It is supporting a status quo that in the short run, at least, shows little prospect of moving toward resolution but which, at the same time, is vulnerable to developments that might degrade it. Washington has said that the actors involved must find a solution but has avoided any suggestion that it would mediate or guarantee any cross-strait agreement. However, the American stance, whether it is "dual deterrence" or "strategic ambiguity," sends a very different message suggesting possible intervention. Once more the United States is the actor in the middle of the cross-strait controversy seeking to maintain equilibrium while balancing relations with both sides.

The more things change...

Notes

CHAPTER 1 AN ISLAND OF UNSETTLED STATUS

1 The next paragraphs on the early history draw from Teng (2004).
2 Terminology referring to the ethnic Chinese who came to the island before the end of World War II is always a problem. In the discussion that follows I will call them "Taiwanese" to distinguish them from the Chinese who came after the war, although strictly speaking only the aborigines are native to Taiwan.
3 This description of the Japanese occupation draws extensively from Lai et al. (1991).

CHAPTER 2 THE COLD WAR IN ASIA AND AFTER

1 These next two sections draw from Goldstein (2000: 5–26) and the sources cited therein.
2 The discussion of the talks that follows draws from Goldstein (2001: 200–37) and the sources cited therein.
3 The analysis in this section has benefited from two excellent essays on the Nixon demarche, by Accinelli (2005: 9–55) and Foot (2005: 90–115).

CHAPTER 3 NORMALIZATION AND NEW PROBLEMS

1 Most of the material cited in this section is from US Department of State (2013a).
2 The reader will recall this is not exactly what the president said, which was that "there would be no more statements made." The other four points were: "We

will not support any Taiwan independence movement; We will use our influence to discourage Japan from moving into Taiwan as our presence diminishes; We will support any peaceful resolution of the Taiwan issue that can be worked out; We seek normalization. (Nixon–HAK suggested the process would be completed by 1976.)" The memo also noted that "Nixon–HAK made two other pledges as well: We will not participate in arrangements that affect Chinese interests without prior consultation. We will reduce our military forces on Taiwan as progress is made toward normalization."

Ten days later, after discussing the question with Secretary of State Cyrus Vance, Brzezinski appeared to reconsider. It would be "wrong," he wrote, to consider these points as constituting "a secret pledge" which would imply "a formal commitment." Brzezinski now considered that they were best seen as "the Nixon administration's point of view," although the Carter administration might choose "to continue to abide" by them in formulating its own approach (US Department of State 2013a: 55).

3 *"I believe there may still be some give on the Chinese position concerning their unilateral statements, should we decide to select the 'arms sales' choice. That is, we may be able to negotiate with the Chinese over the quantity and type of weapons we will sell to Taiwan after normalization, in exchange for some indication of restraint on their part. But this part of the negotiations will have to be handled at the highest levels and done so by indirection"* (US Department of State 2013a: 471).

Ambassador Woodcock reported to the president with a similar conclusion: "To my mind it was expressed clearly that the Chinese accept that 'full commercial relations' include arms sales as necessary. This cannot, of course, be specifically articulated" (US Department of State 2013a: 473). William Gleysteen, deputy assistant secretary for East Asia, doubted that Deng understood that the reference to commercial relations included arms but did see something of a negotiating position in Hua's comments. In general, he considered that "the most important thing that occurred in the discussion of normalization was that the issue of arms sales surfaced at Chinese initiative in a way leaving the door open." Secretary Vance was somewhat skeptical, referring to Hua's statement as "delphic and ambiguous," and expressing the belief that the arms sales issue remained "the trickiest of all and a potentially insurmountable issue." However, he felt that Hua's raising the subject "opens the door sufficiently for us to begin to probe the limits of PRC tolerance" (ibid.: 488 and 494).

4 The next three paragraphs are based on US Department of State (2013a: 648–9) and the document "USLO, Peking 237," a photocopy of which was provided by Alan Romberg from the Carter Library.

5 The following discussion of the Taiwan Relations Act draws from Goldstein and Schriver (2001) and the sources cited therein.

6 For an excellent discussion of the ambiguities and different interpretations of this document, see Bush (2004: 160–75).

7 The following discussion of the nature of the ambiguous rules governing Sino-American relations on the Taiwan issue owes a deep intellectual debt to the conclusion drawn by Romberg in his "The Taiwan tangle" (2006), as well as to our many, many conversations.

CHAPTER 4 THE CHALLENGES OF A DEMOCRATIC TAIWAN

1 The following discussion draws from Newell (1994).

2 Lin (1998) traces the evolution of Lee's policies.

3 This discussion of Lee's visit is based on Suettinger (2003: ch. 6).

4 The following discussion of the 1996 crisis draws from Ross (2000: 87–123), Tyler (1999: 18–43), and Suettinger (2003: ch. 6).

5 The following discussion of the Clinton administration draws from Goldstein and Schriver (2001) and the sources cited therein.

CHAPTER 5 PERIOD OF HIGH DANGER

1 This paragraph and the next three draw on Romberg (2006).

CHAPTER 6 SATISFYING WASHINGTON AND BEIJING

1 A survey found that, in 2014, 34.3 percent of respondents favored a decision later, 25.2 favored indefinitely, 18 percent favored independence later,

7.9 percent favored unification later, 5.9 percent favored independence as soon as possible, and 1.3 percent favored unification as soon as possible (Election Study Center 2015).

2 Except where indicated, the trends described in the next paragraphs are drawn entirely from the surveys conducted by the television station TVBS (http://home.tvbs.com.tw) and Taiwan Indicators Research (www.tisr .com.tw).

CHAPTER 7 ECONOMIC RELATIONS

1 Statistics provided by the mainland and Taiwan differ quite significantly. In the discussion that follows I will use Taiwan customs figures with mainland figures following in parentheses. The source for various numbers is Mainland Affairs Council (2015). For a discussion of the discrepancy between the two statistics, see Rosen and Wang (2011: 7–8). For a rationale for the use of Taiwan statistics, see Tung (2004).

2 This discussion of cross-strait trade and investment has been drawn from Tung Chen-yuan's analysis of its early development (Tung 2002).

3 Statistics in this section are all calculated from figures at the Bureau of Foreign Trade of the Republic of China (Bureau of Foreign Trade 2015).

4 The next three paragraphs draw from Tung (2002: 46–7 and 64–6).

5 The following discussion of Chen Shui-bian's economic policies draws from Tung (2008: 241–57) and Kastner (2009: 60–75).

6 This discussion of direct links is based on Tung (2008: 240–57).

7 The discussion regarding the WTO draws from Chang and Goldstein (2007: 1–42).

8 The following discussion has benefited from three works on this subject: Tung (2002), Kastner (2009), and Tanner (2007).

CHAPTER 8 THE SECURITY DIMENSION

1 The following discussion of Chinese strategy is based on Cliff et al. (2007) and O'Rourke (2014).

2 This discussion of Chen's defense policies draws from Tsang (2008: 259–88).
3 The discussion which follows draws from the blog of Michal Thim regarding defense policy, which can be found at http://taiwan-in-perspective.com.
4 The following discussion of US arms sales draws from Kan (2014).

CONCLUSION

1 This draws from Goldstein (2002). For another approach to understanding the concept of status quo and its future, see Bush (2011).

References

Accinelli, R. (1996) *United States policy toward Taiwan, 1950–1955: Crisis and commitment*. Chapel Hill and London: University of North Carolina Press.

Accinelli, R. (2005) In pursuit of modus vivendi: the Taiwan issue and Sino-American rapprochement, 1969–1972, in W. C. Kirby, R. S. Ross, and L. Gong, eds, *Normalization of U.S.–China relations: an international history*. Cambridge, MA: Harvard University Press, pp. 9–55.

BBC (2000) China ups pressure on Taiwan, March 15. Available at: http://news.bbc.co.uk/2/hi/asia-pacific/678155.stm [accessed 27 Feb 2015].

BBC (2001) Beijing snubs Taiwan's links move, January 3. Available at: http://news.bbc.co.uk/2/hi/asia-pacific/1098950.stm [accessed 27 Feb 2015].

BBC (2005) Text of KMT–Beijing agreement, April 29. Available at: http://news.bbc.co.uk/2/hi/asia-pacific/4498791.stm [accessed 27 Feb 2015].

BBC (2008) Direct China–Taiwan travel agreed, June 13. Available at: http://news.bbc.co.uk/2/hi/asia-pacific/7452097.stm [accessed 28 Feb 2015].

Brown, D. (2001) China–Taiwan relations: wooing Washington, *Comparative Connections*, 3(1). Available at: http://csis.org/files/media/csis/pubs/0101qchina_taiwan.pdf [accessed 27 Feb 2015].

Brown, D. (2006) China–Taiwan relations: missed opportunities, *Comparative Connections*, 8(1). Available at: http://csis.org/files/media/csis/pubs/0601qchina_taiwan.pdf [accessed 28 Feb 2015].

Bureau of Foreign Trade (2015) *Trade statistics*. Available at: http://cus93.trade.gov.tw/ENGLISH/FSCE/ [accessed 28 Feb 2015].

Bush, R. C. (2004) *At cross purposes: U.S.–Taiwan relations since 1942*. Armonk, NY: M. E. Sharpe.

Bush, R. C. (2005) *Untying the knot: making peace in the Taiwan Strait*. Washington, DC: Brookings Institution Press.

Bush, R. C. (2011) Taiwan and East Asian security, *Orbis*, 55(2): 274–89.

Bush, R. C. (2014) *Taiwan and the Trans-Pacific Partnership: the political dimension*. Washington, DC: Brookings Institution, East Asia Policy Paper.

Available at: www.brookings.edu/research/papers/2013/10/03-taiwan-trans-pacific-partnership-bush [accessed 28 Feb 2015].

Cairo Communiqué (1943) December 1. Available at: www.ndl.go.jp/constitution/e/shiryo/01/002_46/002_46tx.html [accessed 5 March 2015].

Chang, C., and Tien, H. (1996) *Taiwan's electoral politics and democratic transition.* Armonk, NY: M. E. Sharpe.

Chang, G. H. (1990) *Friends and enemies: the United States, China, and the Soviet Union, 1948–1972.* Stanford, CA: Stanford University Press.

Chang, J. (2005) Taiwan's policy toward the United States, in W. C. Kirby, R. S. Ross, and L. Gong, eds, *Normalization of U.S.–China relations: an international history.* Cambridge, MA: Harvard University Press.

Chang, J., and Goldstein, S. M. (2007) *Economic reform and cross-strait relations.* Singapore: World Scientific.

Chang, Y. (2003) Lee, Chen make firm commitment to constitution, *Taipei Times,* October 5. Available at: www.taipeitimes.com/News/front/archives/2003/10/05/2003070430 [accessed 27 Feb 2015].

Chen, C., Melachroinos, K. A., and Chang, K. (2010) FDI and local economic development: the case of Taiwanese investment in Kunshan, *European Planning Studies,* 18(2): 213–38.

Chen, S. (2001) *President Chen's national day message,* October 12. Available at: http://taiwaninfo.nat.gov.tw/ct.asp?xItem=18964&CtNode=103&htx_TRCategory=&mp=4 [accessed 27 Feb 2015].

Chien, S. and Zhao, L. (2010) Local economic transition in China: a perspective on Taiwan investment, in J. Wong and Z. Bo, eds, *China's Reform in Global Perspective.* Singapore: World Scientific, pp. 233–60.

Chin, C. (1997) Division of labor across the Taiwan Strait, in B. Naughton, ed., *The China Circle.* Washington, DC: Brookings Institution Press, pp. 164–209.

China Post (2006) Taiwan independence means war for U.S., Zoellick says, May 12. Available at: www.chinapost.com.tw/news/detail.asp?id=82063 [accessed 27 Feb 2015].

China Post (2014) DPP now a strong contender to helm the nation in '16 presidential election, November 30. Available at: www.chinapost.com.tw/taiwan/national/national-news/2014/11/30/423004/DPP-now.htm [accessed 28 Feb 2015].

China–US (1982) *Joint communique between the People's Republic of China and the United States of America,* August 17. Available at: www.china.org.cn/english/china-us/26244.htm [accessed 26 Feb 2015].

Chiu, H. (1973) *China and the question of Taiwan.* New York: Praeger.

Chou, Y., and Nathan, A. J. (1987) Democratizing transition in Taiwan, *Asian Survey,* 27(3): 277–99.

Christensen, T. J. (2007) A strong and moderate Taiwan, speech to US–Taiwan Business Council, September 11. Available at: http://2001-2009.state.gov/p/eap/rls/rm/2007/91979.htm [accessed 27 Feb 2015].

Chu, M., and Ko, S. (2002) Chang gives his approval to passports, *Taipei Times*, January 15. Available at: www.taipeitimes.com/News/front/archives/2002/01/15/0000119980 [accessed 27 Feb 2015].

Chu, Y., and Nathan, A. J. (2007–8) Seizing the opportunity for change in the Taiwan Strait, *Washington Quarterly*, 31(1): 77–91.

Chyan, A. (2014) China–S. Korea FTA to hurt Taiwan. *China Post*, November 11. Available at: www.chinapost.com.tw/taiwan-business/2014/11/11/421528/China-S-Korea.htm [accessed 28 Feb 2015].

Cliff, R., et al. (2007) *Entering the dragon's lair: Chinese antiaccess strategies and their implications for the United States*. Santa Monica, CA: RAND.

CNN (2001) Bush vows "whatever it takes" to defend Taiwan, April 25. Available at: http://edition.cnn.com/2001/ALLPOLITICS/04/25/bush.taiwan.03/ [accessed 27 Feb 2015].

Commentary warns against Li Teng-hui's visit (1995) Foreign Broadcast Information, daily report, FBIS-CHI-95-106, pp. 50–1.

CSIS (Center for Strategic and International Studies) (1994) Taiwan's White Paper on cross-strait relations, July 5. Available at: http://csis.org/files/media/csis/programs/taiwan/timeline/sums/timeline_docs/CSI_19940705.htm [accessed 26 Feb 2015].

Defense News (2014) *Defense news | News about defense programs, business, and technology*. Available at: http://www.defensenews.com [accessed 1 March 2015].

deLisle, J. (2009) *Taiwan in the World Health Assembly: a victory, with limits*. Brookings East Asia Commentary, no. 29. Available at: www.brookings.edu/research/opinions/2009/05/taiwan-delisle [accessed 27 Feb 2015].

DPP (Democratic Progressive Party) (1999) *DPP party convention*. Available at: www.taiwandc.org/nws-9920.htm [accessed 27 Feb 2015].

DPP (Democratic Progressive Party) (2014) *DPP perspective on the cross strait trade in Services Agreement*. Available at: http://english.dpp.org.tw/dpp-perspective-on-the-cross-strait-trade-in-services-agreement/ [accessed 28 Feb 2015].

Dumbaugh, K. (2007) *Underlying strains in Taiwan–U.S. political relations*. Washington, DC: Library of Congress.

Election Study Center (2015a) *Taiwan independence vs. unification with the mainland trend distribution in Taiwan (1992/06~2014/12)*. Available at: http://esc.nccu.edu.tw/course/news.php?Sn=203 [accessed 28 Feb 2015].

Election Study Center (2015b) *Taiwanese/Chinese Identification trend distribution in Taiwan (1992/06~2014/12)*. Available at: http://esc.nccu.edu.tw/course/news.php?Sn=203 [accessed 28 Feb 2015].

ESRC (US–China Economic and Security Review Commission) (2014) *Hearing: China's military modernization and its implications for the United States*, January 30. Available at: http://origin.www.uscc.gov/Hearings/hearing-china%E2%80%99s-military-modernization-and-its-implications-united-states [accessed 1 March 2015].

Evans, T. (2010) *Ma: Taiwan won't ask U.S. to fight China*, May 1. Available at: http://edition.cnn.com/2010/WORLD/asiapcf/04/30/taiwan.china.us/index.html [accessed 1 March 2015].

Fatal Politics (1971) *Polo I briefing book: Kissinger's secret trip to China, July 1971*. Available at: www.scribd.com/doc/58767074/Polo-I-Briefing-Book-Kissinger-s-Secret-Trip-to-China-July-1971 [accessed 22 Feb 2015].

Federation of American Scientists (2000) *Full text of President Chen Shui-bian's inaugural speech*, May 20. Available at: http://fas.org/news/taiwan/2000/e-05-20-00-8.htm [accessed 27 Feb 2015].

Feldman, H. (2001) Taiwan, arms sales, and the Reagan assurances, *American Asian Review*, 19(3), pp. 75–101.

Foot, R. (2005) Prizes won, opportunities lost: the U.S. normalization of relations with China, 1972–1979, in W. C. Kirby, R. S. Ross, and L. Gong, eds, *Normalization of U.S.–China relations: an international history*. Cambridge, MA: Harvard University Press, pp. 90–115.

Garver, J. W. (1997) *The Sino-American alliance: Nationalist China and American Cold War strategy in Asia*. Armonk, NY: M. E. Sharpe.

Gilley, B. (2010) Not so dire straits: how the Finlandization of Taiwan benefits U.S. security, *Foreign Affairs*, January–February. Available at: www.foreignaffairs.com/articles/65901/bruce-gilley/not-so-dire-straits [accessed 28 Feb 2015].

Goldstein, S. M. (1999) Terms of engagement: Taiwan's mainland policy, in A. Johnston and R. Ross, eds, *Engaging China: the management of an emerging power*. New York and London: Routledge, pp. 57–86.

Goldstein, S. M. (2000) *The United States and the Republic of China, 1949–1978: suspicious allies*, APARC Working Paper, Stanford University.

Goldstein, S. M. (2001) Dialogue of the deaf? Sino-American ambassadorial-level talks, 1955–1970, in R. S. Ross and J. Changbin, eds, *Re-examining the Cold War: U.S.–China diplomacy, 1954–1973*. Cambridge, MA: Harvard University Press, pp. 200–37.

Goldstein, S. M. (2002) The Taiwan Strait: a continuing status quo of deadlock?, *Cambridge Review of International Affairs*, 15(1): 85–94.

Goldstein, S. M. (2007) The cross-strait talks of 1993 – the rest of the story: domestic politics and Taiwan's mainland policy, *Journal of Contemporary China*, 6(15): 259–85.

Goldstein, S. M. (2008) *Presidential politics in Taiwan: the administration of Chen Shui-bian*. Norwalk, CT: Eastbridge.

Goldstein, S. M., and Schriver, R. (2001) An uncertain relationship: the United States, Taiwan and the Taiwan Relations Act, in R. L. Edmonds and S. M. Goldstein, eds, *Taiwan in the Twentieth Century: A Retrospective View*. New York and London: Cambridge University Press, pp. 147–72.

Harding, H. (1992) *A fragile relationship: the United States and China since 1972*. Washington, DC: Brookings Institution Press.

House Committee on Foreign Affairs (2013) *H.R. 419, Taiwan Policy Act of 2013*. Available at: http://foreignaffairs.house.gov/bill/hr-419-taiwan-policy -act-2013 [accessed 1 March 2015].

Hsiau, A. (2000) *Contemporary Taiwanese cultural nationalism*. London: Routledge.

Hsieh, P. L. (2011)The China–Taiwan ECFA, geopolitical dimensions and WTO law, *Journal of International Economic Law*, 14(1): 121–56.

Hung, J. (2014) Tsai Ing-wen's "Taiwan consensus," *China Post*, July 28. Available at: www.chinapost.com.tw/commentary/the-china-post/joe-hung/2014/07 /28/413400/Tsai-Ing-wens.htm [accessed 28 Feb 2015].

Kan, S. (2011) *China/Taiwan: evolution of the "one China" policy: key statements from Washington, Beijing, and Taipei*. Washington, DC: Library of Congress.

Kan, S. (2014) *Taiwan: major U.S. arms sales since 1990*. Congressional Research Service. Available at: http://fas.org/sgp/crs/weapons/RL30957.pdf [accessed 1 March 2015].

Kastner, S. L. (2009) *Political conflict and economic interdependence across the Taiwan Strait and beyond*. Stanford, CA: Stanford University Press.

Kerr, G. H. (1965) *Formosa betrayed*. Boston: Houghton Mifflin.

KMT (2015) *TAO: 1992 Consensus is anchor of cross-strait relations*, January 21. Available at: www.kmt.org.tw/english/page.aspx?type=article&mnum=112 &anum=15706 [accessed 28 Feb 2015].

Kwan, D. (2000) "No more room for tomfoolery," *South China Morning Post*, May 22.

Lai, Z., Myers, R. H., and Wei, E. (1991) *A tragic beginning: the Taiwan uprising of February 28, 1947*. Stanford, CA: Stanford University Press.

Lee, C. (2006) Cluster adaptability across sector and border: the case of Taiwan's information technology industry. PhD dissertation, University of California, Berkeley.

Lee, T. (1995) Always in my heart, lecture given at Cornell University, June 9. Available at: http://csis.org/files/media/csis/programs/taiwan/timeline/ sums/timeline_docs/CSI_19950609.htm [accessed 26 Feb 2015].

Lee, T. (1996) *Inaugural address*, May 20. Available at: http://newcongress.yam. org.tw/taiwan_sino/leespeec.html [accessed 27 Feb 2015].

Lee, T. (1999) *President Lee Teng-hui's responses to questions submitted by Deutsche Welle*, July 9. Available at: http://fas.org/news/taiwan/1999/0709.htm [accessed 27 Feb 2015].

Leng, T. (1996) *The Taiwan–China connection: democracy and development across the Taiwan Straits*. Boulder, CO: Westview Press.

Li's "lining" means creating "two Chinas" (1995) Foreign Broadcast Information, daily report, CHI-95-174, pp. 54–8.

Li's "political scheme" (1995) Foreign Broadcast Information, daily report, CHI-95-109, pp. 82–7.

Liaowang views "independent Taiwan" (1990) Foreign Broadcast Information, daily report, CHI-90-052-S.

Lin, C. (1998) *Paths to democracy: Taiwan in comparative perspective*. New Haven, CT: Yale University Press.

Lin, C. (2000) Chen wants to return to 1992 stance, *Taipei Times*, August 1. Available at: www.taipeitimes.com/News/front/archives/2000/08/01/0000 045964 [accessed 27 Feb 2015].

Liu, K. (2001) Chen searches for common ground, *Taipei Times*, January 5. Available at: www.taipeitimes.com/News/editorials/archives/2001/01/05/00000 68530 [accessed 27 Feb 2015].

Lo, V. (2000) Chen's remarks nothing new: MAC, *Taipei Times*, June 29. Available at: www.taipeitimes.com/News/front/archives/2000/06/29/0000041 839 [accessed 27 Feb 2015].

Luo, Q. (1999) Cross-strait economic relations since 1999: economic dynamism and political fragility, in G. Wang and J. Wong, ed., *China: Two Decades of Reform and Change*. Singapore: Singapore University Press.

Mainland Affairs Council (1991) *Guidelines for national unification*. Available at: https://law.wustl.edu/Chinalaw/twguide.html [accessed 26 Feb 2015].

Mainland Affairs Council (2014a) Public opinion on current cross-strait relations: statistic charts. Available at: www.mac.gov.tw/ct.asp?xItem=109563&c tNode=7630&mp=3 [accessed 27 Feb 2015].

Mainland Affairs Council (2014b) *Mainstream public supports MAC in continued promotion of institutionalized cross-strait negotiations Mainland Affairs Council*, December 25. Available at: www.mac.gov.tw/ct.asp?xItem=110822&ctNode =6337&mp=3 [accessed 28 Feb 2015].

Mainland Affairs Council (2015) *Jingmao jiaoliu tongji* [Economy and trade flow statistics]. Available at: www.mac.gov.tw/np.asp?ctNode=5712&mp=1 [accessed 28 Feb 2015].

Mann, J. (1999) *About face: a history of America's curious relationship with China from Nixon to Clinton*. New York: Alfred Knopf.

Meltzer, J. (2011) *The Trans-Pacific Partnership: its economic and strategic implications*. Brookings Institution. Available at: www.brookings.edu/research/opinions/2011/09/30-trans-pacific-partnership-meltzer [accessed 28 Feb 2015].

Ministry of National Defense (2009) *Quadrennial defense review*. Available at: www.us-taiwan.org/reports/taiwan_2009_qdr.pdf [accessed 1 March 2015].

Ministry of National Defense (2013) *Quadrennial defense review*. Available at: http://qdr.mnd.gov.tw/file/2013QDR-en.pdf [accessed 1 March 2015].

Murray, W. S. (2008) Revisiting Taiwan's defense strategy, *Naval War College Review*, 61(3): 13–38.

National Archives & Records Administration (1945) Japan surrenders. Available at: www.archives.gov/exhibits/featured_documents/japanese_surrender_document [accessed 15 April 2005].

National People's Congress (2005) *Anti-Secession Law*, adopted March 14. Available at: www.china.org.cn/english/2005lh/122724.htm [accessed 28 Feb 2015].

Newell, P. (1994) *The transition to the transition toward democracy in Taiwan: political change in the Chiang Ching-kuo era, 1971–1986*. PhD dissertation, Georgetown University.

O'Rourke, R. (2014) *China naval modernization: implications for U.S. Navy capabilities*. Washington, DC: Library of Congress.

Potsdam Declaration (1945) Proclamation defining terms for Japanese surrender, July 26. Available at: www.ndl.go.jp/constitution/e/etc/c06.html [accessed 5 March 2015].

Renmin Ribao views Taiwan "one China" stance (1994) Foreign Broadcast Information, daily report, CHI-94-167, pp. 92–5.

Reuters (2013) China's Xi says political solution for Taiwan can't wait forever. Available at www.reuters.com/article/2013/10/06/us-asia-apec-china-taiwan-idUSBRE99503Q20131006 [accessed April 15, 2015].

Risenhoover, P. M. (2007) *Official US State Department Memorandum on Legal Status of Taiwan: memorandum, July 13, 1971*. Available at: http://chinasupport.blogspot.com/2007/03/official-us-state-department-memorandum.html [accessed 5 March 2015].

Romberg, A. D. (2003) *Rein in at the brink of the precipice: American policy toward Taiwan and US–PRC relations*. Washington, DC: Henry L. Stimson Center.

Romberg, A. D. (2006) The Taiwan tangle, *China Leadership Monitor*, no. 18, July. Available at: www.hoover.org/research/taiwan-tangle [accessed 27 Feb 2015].

Romberg, A. D. (2009) Cross-strait relations: first the easy, now the hard, *China Leadership Monitor*, no. 28, May. Available at: www.hoover.org/research/first-easy-now-hard [accessed 27 Feb 2015].

Romberg, A. D. (2010) 2010: the winter of PRC discontent, *China Leadership Monitor*, no. 31, February. Available at: www.hoover.org/research/2010-winter-prc-discontent [accessed 28 Feb 2015].

Romberg, A. D. (2014a) From generation to generation: advancing cross-strait relations, *China Leadership Monitor*, no. 43, February. Available at: www.hoover.org/research/generation-generation-advancing-cross-strait-relations [accessed 28 Feb 2015].

Romberg, A. D. (2014b) Sunshine heats up Taiwan politics, affects PRC tactics, *China Leadership Monitor*, no. 44, June. Available at: www.hoover.org/publications/china-leadership-monitor/china-leadership-monitor-summer-2014 [accessed 28 Feb 2015].

Rosen, D. H., and Wang, Z. (2011) *The implications of China–Taiwan economic liberalization*. Washington, DC: Peterson Institute for International Economics.

Ross, R. S. (2000) The 1995–1996 Taiwan Strait confrontation: coercion, credibility, and use of force, *International Security*, 25(2): 87–123.

Rothkopf, D. (2010) Can the U.S. afford to continue supporting Taiwan?, *Foreign Policy*, February 1. Available at: http://rothkopf.foreignpolicy.com/posts/2010/02/01/can_the_us_afford_to_continue_supporting_taiwan [accessed 28 Feb 2015].

Roundup reviews 1989 political situation (1990) Foreign Broadcast Information, daily report, FBIS-CHI-025.

Shih, E. (2004) *The conduct of U.S.–Taiwan relations, 2000–2004*. CNAPS Working Paper, Brookings Institution, October.

Shih, H. (2012) US beef is political hostage: AIT boss, *Taipei Times*, June 28. Available at: www.taipeitimes.com/News/taiwan/print/2012/06/28/2003536452 [accessed 28 Feb 2015].

Shlapak, D. A. (2009) *A question of balance: political context and military aspects of the China–Taiwan dispute*. Santa Monica, CA: RAND.

Snyder, C. (2007) ROC statehood undecided: US official, *Taipei Times*, September 1. Available at: www.taipeitimes.com/News/front/archives/2007/09/01/2003376690 [accessed 27 Feb 2015].

Solomon, R. H. (1999) *Chinese negotiating behavior: pursuing interests through "old friends"*. Washington, DC: United States Institute of Peace Press.

Su, C. (2009) *Taiwan's relations with mainland China: a tail wagging two dogs*. London and New York: Routledge.

Suettinger, R. (2003) *Beyond Tiananmen: the politics of U.S.–China relations, 1989–2000*. Washington, DC: Brookings Institution Press.

Sutter, R. (2002) The Bush administration and U.S. China policy debate: reasons for optimism, *Issues and Studies*, 38(2): 1–30.

Sutter, R., Brown, M. E., Adamson, T. J. A., Mochizuki, M. M., and Ollapally, D. (2013) *Balancing acts: the U.S. rebalance and Asia–Pacific stability*. Washington, DC: Sigur Center for Asian Studies.

Swaine, M. D., and Mulvenon, J. C. (2001) *Taiwan's foreign and defense policies*. Santa Monica, CA: RAND.

Taipei Times (2008) Hu Jintao putting "deep" thought into cross-strait issues, *Taipei Times*, April 14. Available at: www.taipeitimes.com/News/taiwan/archives/2008/04/14/2003409209 [accessed 28 Feb 2015].

Taiwan Affairs Office (2000) *The one-China principle and the Taiwan issue*. Available at: www.china.org.cn/english/taiwan/7956.htm [accessed 27 Feb 2015].

Taiwan Affairs Office (2008) *President Hu offers six proposals for peaceful development of cross-strait relationship*, December 31. Available at: www.gwytb.gov.cn/en/Special/Hu/201103/t20110322_1794707.htm [accessed 28 Feb 2015].

Taiwan Affairs Office (2011) *White papers on Taiwan issue*. Available at: www.gwytb.gov.cn/en/Special/WhitePapers/201103/t20110316_1789217.htm [accessed 28 Feb 2015].

Taiwan Communiqué (2002) no. 102, September. Available at: www.taiwandc.org/twcom/102-no1.htm [accessed 27 Feb 2015].

Taiwan Documents Project (1952) Treaty of peace with Japan. Available at www.taiwandocuments.org/sanfrancisco01.htm [accessed 15 April 2015].

Taiwan in Perspective (2014) *Taiwan in perspective*. Available at: http://taiwan-in-perspective.com/ [accessed 1 March 2015].

Tanner, M. S. (2007) *Chinese economic coercion against Taiwan*. Santa Monica, CA: RAND.

Teng, E. (2004) *Taiwan's imagined geography*. Cambridge, MA: Harvard University Press.

Tien, H. (1989) *The great transition: political and social change in the Republic of China*. Stanford, CA: Hoover Institution Press, Stanford University.

Truman, H. S. (1950a) The president's news conference, January 5. Available at: www.presidency.ucsb.edu/ws/index.php?pid=13678 [accessed April 15, 2015]

Truman, H. S. (1950b) Statement by the president on the situation in Korea. Available at: www.presidency.ucsb.edu/ws/index.php?pid=13538&st=Korea &st1= [accessed April15, 2015].

Tsang, S. (2008) Taiwan's changing security environment, in S. M. Goldstein and J. Chang, eds, *Presidential politics in Taiwan: the administration of Chen Shui-bian*. Norwalk, CT: EastBridge, pp. 259–88.

Tucker, N. B. (2009) *Strait talk: United States–Taiwan relations and the crisis with China*. Cambridge, MA: Harvard University Press.

Tung, C. (2002) *China's economic leverage and Taiwan's security concerns with respect to cross-strait economic relations*. PhD dissertation, Johns Hopkins University.

Tung, C. (2004) *Economic relations between Taiwan and China*, UNISCI Discussion Papers, no. 4. Available at: www.redalyc.org/pdf/767/76712465003.pdf [accessed 28 Feb 2015].

Tung, C. (2008) The evolution and assessment of cross-strait economic policy in the first term of the Chen Shui-bian administration, in S. M. Goldstein and J. Chang, eds, *Presidential politics in Taiwan: the administration of Chen Shui-bian*. Norwalk, CT: EastBridge, pp. 240–57.

TVBS Poll Center (2013) *Ma-xi huiyi yu guozu rentong mindiao* [Opinion poll on a Ma-Xi meeting and national identity]. Available at: http://home.tvbs .com.tw/static/FILE_DB/PCH/201311/20131106112520608.pdf [accessed 28 Feb 2015].

Tyler, P. (1999) *A great wall: six presidents and China*. New York: Public Affairs.

US Department of Defense (2013) *Air-Sea Battle*. Available at: www.defense. gov/pubs/ASB-ConceptImplementation-Summary-May-2013.pdf [accessed 1 March 2015].

US Department of Defense (2014) *Annual report to Congress: military and security developments involving the People's Republic of China 2014*. Available at: www .defense.gov/pubs/2014_DoD_China_Report.pdf [accessed 28 Feb 2015].

US Department of State (2003) President Bush and Premier Wen Jiabao remarks to the press, December 9. Available at: http://2001-2009.state.gov/p/eap/ rls/rm/2003/27184.htm [accessed 27 Feb 2015].

US Department of State (2006) *Foreign Relations of the United States, 1969–1976*, Vol. XVII: China, 1969–1972. Available at: http://history.state.gov/ historicaldocuments/frus1969-76v17 [accessed 22 Feb 2015].

US Department of State (2010) Daily press briefing, June 29. Available at: www .state.gov/r/pa/prs/dpb/2010/06/143757.htm [accessed 28 Feb 2015].

US Department of State (2013a) *Foreign Relations of the United States, 1977–1980*, Vol. XIII: China. Available at: https://history.state.gov/ historicaldocuments/frus1977-80v13 [accessed 22 Feb 2015].

US Department of State (2013b) Taiwan's participation in the International Civil Aviation Organization (ICAO), press statement, September 24.

Available at: www.state.gov/r/pa/prs/ps/2013/09/214658.htm [accessed 27 Feb 2015].

Wachman, A. (2007) *Why Taiwan? Geostrategic rationales for China's territorial integrity.* Stanford, CA: Stanford University Press.

Wang, C. (2012) Frank Hsieh confirms visit to China, *Taipei Times*, October 2. Available at: www.taipeitimes.com/News/front/archives/2012/10/02/2003 544157 [accessed 28 Feb 2015].

Wang, X. (2004) *Chen Yi yu erh erh ba shijian [Chen Yi and the 2/28 incident].* Taipei: Haixiaxueshu.

Wen, Q. (2009) "One country, two systems": the best way to peaceful reunification, *Beijing Review*, May 26. Available at: www.bjreview.com.cn/nation/txt/2009-05/26/content_197568.htm [accessed 26 Feb 2015].

Xing, Z. (2002) Beijing fires strong warning, *China Daily*, August 6. Available at: www.chinadaily.com.cn/en/doc/2002-08/06/content_130856.htm [accessed 27 Feb 2015].

Xinhuanet (2002) *Acceptance of 1992 Consensus essential to reopen negotiations: signed article.* Available at: http://news.xinhuanet.com/english/2002-05/11/content_389001.htm [accessed 6 July 2014].

Xinhuanet (2005) *Four-point guidelines on cross-straits relations set forth by President Hu.* Available at: http://news.xinhuanet.com/english/2005-03/04/content_2651270.htm [accessed 27 Feb 2015].

Xinhuanet (2012) *Full text of Hu Jintao's report at 18th Party Congress*, November 17. Available at: http://news.xinhuanet.com/english/special/18cpcnc/2012-11/17/c_131981259_11.htm [accessed 28 Feb 2015].

Xinhuanet (2013) *Mainland, Taiwan negotiators sign service trade agreement*, June 21. Available at: http://news.xinhuanet.com/english/china/2013-06/21/c_132475046.htm [accessed 28 Feb 2015].

Yang, P. (2008) Cross-strait relations in the first Chen administration, in S. M. Goldstein and J. Chang, eds, *Presidential politics in Taiwan: the administration of Chen Shui-bian.* Norwalk, CT: EastBridge, pp. 203–28.

Index